ALL THE MATH THAT'S FIT TO PRINT

Articles from The Manchester Guardian

ALL THE MATH

HAT'S FIT TO PRINT

Articles from *The Manchester Guardian*

KEITH DEVLIN

HE MATHEMATICAL ASSOCIATION OF AMERICA

1529 EIGHTEENTH STREET, N.W.   WASHINGTON, DC 20036

Sources for Figures:

Figures 1, 9, 14, 15, 17, 19, 20, 22, 23, 35, and 36: Keith Devlin, *Mathematics: The New Golden Age,*
   London: Penguin Books, 1988, pp. 2, 170, 266, 274, 79, 78, 93, 76, 90, 231, and 260, respectively.
Figures 10 and 11: E. A. Abbot, *Flatland,* Dover, 1952, pp. ii and 81, respectively.
Figure 13 and 21: H. E. Huntley, *The Divine Proportion*, New York, Dover, 1970, pp. 165 and 63,
respectively.
Figures 24 and 25: F. J. Bloore and H. R. Morton, Advice on hanging pictures, *American Mathematical
   Monthly* 92 (May 1985) pp. 310 and 313, respectively.
Figure 26: S. Ulam, *Science, Computers, and People*, Boston: Birkhäuser, 1986, p. 95.
Figure 32: D. Rumelhart and J. McClelland, *Parallel Distributed Processing*, Cambridge, MA: MIT Press,
   1986, p. 22.
Figure 33: Peter M. Maurer, A rose is a rose, *American Mathematical Monthly* 94 (August-September 1987)
   p. 633.

Figures 3, 4, and 5 are reprinted with permission from Arno Peters, *The New Cartography*, New York:
Friendship Press, 1983, pp. 100, 124, and 106, respectively. © 1983 Universitäsverlag Carinthia.

Figure 12 is reprinted with permission from A. K. Dewdney, *The Planiverse*, London: Pan Books, 1984, p.
40. © 1984 Pan Books.

© *1994 by*
The Mathematical Association of America (Incorporated)
Library of Congress Catalog Card Number 94-77346

ISBN 0-88385-515-1

Printed in the United States of America

Current Printing (last digit):
10 9 8 7 6 5 4 3 2 1

SPECTRUM SERIES

The Spectrum Series of the Mathematical Association of America was so named to reflect its purpose: to publish a broad range of books including biographies, accessible expositions of old or new mathematical ideas, reprints and revisions of excellent out-of-print books, popular works, and other monographs of high interest that will appeal to a broad range of readers, including students and teachers of mathematics, mathematical amateurs, and researchers.

Mathematical Association of America
1529 Eighteenth Street, NW
Washington, DC 20036
800-331-1MAA   FAX 202-265-2384

Preface

I conceive that these things, king Gelon, will appear incredible to the great majority of people who have not studied mathematics. But to those ... [that have] ... I thought the subject would be not inappropriate for your attention.
   —Archimedes, at the close of his article *The Sand Reckoner*, in which he calculates that the number of grains of sand required to completely fill the entire visible universe is:
   100,000,000,000,000,000,000,000,000,000,000,000,000,000,000,000,000,000

What kind of book is this?

This is not the kind of book you read cover to cover. It is a book for delving. All the chapters are short (between 600 and 1200 words), all are self-contained, and all have appeared in the British national press. An ideal place to peruse its contents would be the smallest room in the house. Leave it there for all to glance through. There is something for everyone. The mathematical topics range from simple puzzles to deep results and open problems, such as Faltings Theorem and the Riemann Conjecture.

Where did it come from?

Between 1983 and 1989, I wrote a twice-monthly column on mathematics and computing in the English national daily newspaper *The Guardian* (known in the United States as *The Manchester Guardian*). Since relocating to the United States in 1989, I have contributed articles on a less regular basis. This book is a compilation of most of those articles. The articles are presented chronologically, year by year. In preparing this compilation, a few minor changes have been made. I have deleted discussions of newly-published books or mathematically-based products that are no longer available. And I have added a number of cross-references and updates. But for the most part, the articles are exactly as they first appeared in *The Guardian*. For convenience, I have added a topic index, by chapter.

Other titles this book might have had?

Micromaths. For most of its life, my *Guardian* column went under the name "Micromaths." In 1984, Macmillan brought out a small anthology of some of the early columns under the same name. (It is no longer in print.) So I could have used that title. But, having lived with this name for a decade of newspaper writing, I was ready for something new.

Bits and PC's. This was one of the names I considered for the *Guardian* column when it began. It has a nice, snappy, airport-departure-lounge-bookshop ring to it. And it is a very cute pun. It would probably help sales enormously. But it's not very British, is it, and though I now live and work in the United States, I feel I owe my former *Guardian* editors and readers at least a title they can live with.

Pi and Chips. This is clearly the British equivalent of the above. As a Yorkshireman, for whom pie and chips is an ethnic dish, it is my own favorite title, but a sure-fire flop for the American market. My *Guardian* editor didn't like it either when I suggested calling the column by this name.

The Guardian Math Book. An accurate title. But too predictable, like *The Guardian Recipe Book* or *The Guardian Book of Crossword Puzzles.* Readers of this book deserve better than that. So, if the *New York Times* can proclaim proudly that it publishes "all the news that's fit to print," then this book can have a similar title.

Who is the intended audience?

Publishers always ask prospective authors this question. In this case, I know the answer: anyone who regularly reads a serious newspaper and has some interest in matters scientific, mathematical, or who is just plain curious.

From the mail I received, I know that the readers of the Micromaths column were a varied bunch. They ranged from students at schools in their early teens (occasionally even younger!), through to retired people in their nineties (often the ones who best succeed in cracking the brain-teasers I occasionally included in my articles); from prison inmates through to executives in the computer industry (not necessarily separate classifications, I suppose); from truckers to schoolteachers; both men and women.

Mathematics knows no barriers except that of interest, and even those who profess no interest can, when presented with the right bait, be suitably hooked.

Do you need to be "good at math" to read this book? Not at all. If you were interested enough to pick it up and open its pages, you should have no trouble reading the whole thing.

There is no overall theme. In writing the Micromaths column, I simply wrote about current events in mathematics, or else what interested me at the time. So long as I could say what I wanted in 600–1200 words or so, and so long as it had broad enough interest, my *Guardian* editor (initially Tim Radford, then Jack Schofield) would print it.

Mathematical news? Are you kidding?

For those people who think that nothing of interest has happened in mathematics since the time of Pythagoras, I have news for you. There are recent developments in mathematics! It is not a dry old subject, all worked out centuries ago. It is a living human enterprise, both a science and an art, that reflects certain aspects of we, its creators. That aspect too should come through in the following pages. For along with the puzzles and recreations are up-to-the-minute news articles—at least they were up-to-the-minute when they first appeared in the pages of *The Guardian,* which is, after all, a *news* paper.

What kind of newspaper would have a mathematics column?

What is this newspaper *The Guardian,* that would dare to publish a regular column on mathematics? (And remember, the column began way back in 1983, before mathematics was regarded as having *any* news appeal at all, let alone warranting regular coverage.)

In Britain, at least, the caricature image of the typical *"Guardian* reader" is of a middle class, college educated, politically left-of-center, jeans-and-sweater, male or female in their late thirties or early forties, who reads their *Guardian* over the breakfast table.

My mailbag alone belied this popular image, but *Guardian* writers quite like it (though they will not admit it). Certainly, it is a newspaper whose readership is both sophisticated and fiercely loyal. And enough of these readers seemed to like their morning paper to come with an occasional dose of mathematics, and that provided the origins to this book.

Who is this guy Devlin anyway?

Who am I? When I am giving talks, people often introduce me by saying that, in addition to my journalistic and television work, I am also a "serious mathematician." What they mean by this (at least, this is what I *hope* they mean) is that I am a professional, research mathematician and mathematics educator. I have been in the mathematics profession for over twenty years now. I obtained my Ph.D. degree in mathematics from the University of Bristol in 1971. Since then, I have written some fourteen books, ranging from advanced textbooks to the popular account *Mathematics: The New Golden Age,* published by Penguin in 1988, and over 50 research articles. In addition to my research work, I have been involved in the production of a number of television programs dealing with mathematical themes (including the 1985 special *A Mathematical Mystery Tour*), and one radio play, *A Million to One Chance,* based loosely on probability theory, broadcast by the BBC in 1985. And of course, between 1983 and 1989 I wrote the twice-monthly Micromaths column in *The Guardian.*

The educational ravages of Thatcherism having forced me from my native British shores in 1987, I am currently Dean of the School of Science and Professor of Mathematics at Saint Mary's College of California, from where I edit *FOCUS,* the newsletter of the Mathematical Association of America, and the "Computers and Mathematics" column of the American Mathematical Society *Notices.*

Keith Devlin,
Moraga, California
March 1994

Tribute

A wizard who weaves magic into mathematics

The following article appeared on October 25, 1984. It is presented here as a tribute to the master of all mathematics writers, Martin Gardner, who served as an example to all of us who tried to emulate him.

What is the secret of good mathematics writing? When I was first thinking about trying to persuade the *Guardian* to include the occasional article on mathematics among its regular science coverage, and was faced with the task of providing some samples of the kind of thing I had in mind, I did what anyone else in a similar position would have done: I took a long, hard look at the work of the undisputed master of the art, Martin Gardner.

Over the 30 or so years that his column ran in *Scientific American,* Gardner built up an enthusiastic readership, which included practically every professional mathematician that I have met, and no doubt the countless many I have not. There must be a great many professionals whose own interest in the subject was awakened by his rare gift of being able to put across to the outsider the deepest developments in an extremely difficult and abstract subject, and to convey the enjoyment and excitement that lies within mathematics proper.

Though he had no formal training in mathematics, through his column he corresponded with many of the world's leading mathematicians, and indeed it was on occasion as a direct result of what he wrote that mathematics took the turns it did.

There are certain results which I suspect would not have been obtained without his writing about problems. (Results on "Tilings" come to mind, but there are other examples.) Indeed, in 1981, in honor (belatedly) of his 65th birthday, Wadsworth Publishers brought out a special volume entitled *The Mathematical Gardner.* Edited by David Klarner, this volume contained articles written for Gardner by many of the world's leading mathematicians.

Last Sunday, Gardner celebrated his 70th birthday, and it is fitting that on October 18, Viking Books brought out his book *Mathematical Magic Show.* This is one of several updated collections of his articles from *Scientific American.*

If you have never before ready any of Gardner's stuff, now is your chance. His first chapter, on *Nothing* (as seen by the mathematician) is one of the funniest, and at the same time most illuminating pieces he has ever written. It begins with the marvellous quotation from P.L. Heath:

"Nobody seems to know how to deal with it. (He would, of course.)"

Another quotation which I cannot resist passing on is the following:

John Cage's 4′ 33″ is a piano composition that calls for four minutes and thirty-three seconds of total silence as the player sits frozen on the piano stool. The duration of the silence is 273 seconds. This corresponds, Cage has explained to 273 degrees centigrade, or absolute zero, the temperature at which all molecular motion quietly stops. I have not heard 4′ 33″ performed, but friends who have tell me it is Cage's finest composition.

Anyone who can weave items like that into an article on mathematics should be read and read and read.

I could go on quoting from the book, and indeed several of his other books. I have read them all several times over. Indeed, when I was first asked to write a regular column on mathematics for this page just over a year ago, I did what anyone else in my position would have done. I ran off and re-read all of the Martin Gardner books I could lay my hands on to remind myself how it was done. If this column has achieved any measure of success, it is in no small part due to Gardner's influence.

That brings to mind the occasion some years ago when I met a young colleague in Oxford. (Asides like this are a favorite Gardner device.) Having obtained his PhD degree, but unable to secure a university position, he had found temporary employment writing articles on mathematics for an encyclopedia. Knowing that his knowledge of mathematics was, like anyone engaged in highly specialized research, rather narrow in its scope, I asked him how he was able to cover the many diverse topics he had to write on. "That's easy," he said, "I just look it up in an encyclopedia."

I'll finish by passing on two mathematical problems from Gardner's book. First, what is the largest number of proper divisors that a number of four or fewer digits can have, and what numbers have this many divisors? Second, what are the four numbers equal to the sum of the factorials of their digits? Two solutions are obvious. A third can be found with a few moments doodling with paper and pencil. The fourth requires (I would think!) a computer.

Contents

1985

1986

1987

1990 onwards

- *Time*'s Men of the Year:

    Ronald Reagan & Yuri Andropov

- Sally Takes a Ride

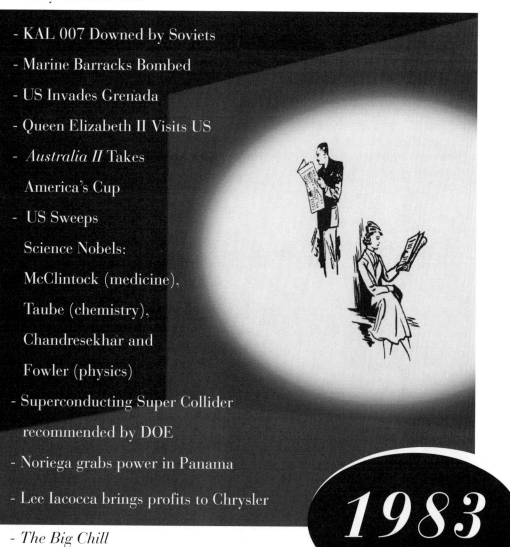

- KAL 007 Downed by Soviets

- Marine Barracks Bombed

- US Invades Grenada

- Queen Elizabeth II Visits US

- *Australia II* Takes
  America's Cup

- US Sweeps
  Science Nobels:
  McClintock (medicine),
  Taube (chemistry),
  Chandresekhar and
  Fowler (physics)

- Superconducting Super Collider
  recommended by DOE

- Noriega grabs power in Panama

- Lee Iacocca brings profits to Chrysler

1983

- *The Big Chill*

- *Flashdance*

- Michael Jackson's *Thriller*

1

The biggest prime number in the world

May 12, 1983

A prime number is any whole number which can only be divided (without recourse to remainders or fractions) by the numbers 1 and itself. (For example, 2, 3, 5, 7, 11 are all primes.) Although the prime numbers have been studied by mathematicians (both professional and recreational) since ancient times, it is only over the last few years that interest has been shown from other quarters.

Recent developments in cryptology (the science of making and breaking secret codes) have tended to involve more and more aspects of the branch of mathematics known as number theory, and in particular the properties of prime numbers. Not surprisingly therefore, large communications and data organizations such as IBM and AT&T now provide extensive funding for research into prime numbers, and less academic agencies such as America's National Security Agency (NSA) are also highly involved in such matters. So it is unlikely that only a handful of ivory-towered mathematicians will show interest in the recent announcement that a new prime has been discovered, a prime immeasurably larger than any known beforehand.

So large is this number that it would be pointless trying to represent it in the way numbers are usually expressed, using a string of the digits 0 to 9. If the editor of the *Guardian* were to decide to print this number in the normal way, using regular sized type, with no headlines, advertisements, or pictures in the way, the number would take up just over an entire half page of the newspaper.

Fortunately, mathematicians have a special notation for describing numbers of this magnitude. Using this notation the number in question looks quite tame. It is $2^{86243} - 1$. That this number is prime was discovered by David Slowinski of the USA. As you might imagine, he had more than a £5 pocket calculator to help him with his calculation. In fact he made use of the world's most powerful computer, the giant Cray-1 machine at the Cray Research Laboratories. Even with this incredible computing power, it took the machine 1 hour 3 min and 22 sec simply to check that the above number is indeed a prime. Months of computing were required to find this number in the first place.

It is not hard to explain what the notation used above means. To obtain Slowinski's number you take the number 2 and multiply it by itself 86,243 times, and then, as a final fillip, you subtract 1. The result is a number with precisely 25,962 digits when written out in the normal way. How can we begin to comprehend the size of such a monster? To get some idea, let's look at the apparently insignificant number 2^{64}. This can be visualized as follows. Imagine an ordinary chessboard. If we number the squares on this chessboard starting in the top left-hand corner and proceeding row by row down to the bottom right-hand corner, using the numbers 1, 2, 3 and so on, the last square we number will get the number 64.

Now imagine that we start putting ten pence pieces on the squares of the chessboard. On square number 1 we put 2 tenpence pieces, on square 2 we put 4, on square 3 we put 8, and so on, on each successive square putting exactly twice as many coins as on the previous one.

On the last square we will form a pile of exactly 2^{64} ten-pence pieces. How high do you think this pile will be? Six feet? Fifty feet? More? Wait for it. The pile will be about 37 million million kilometers high! So the pile would stretch way beyond the moon (a mere 400,000 kilometers away) and the sun (150

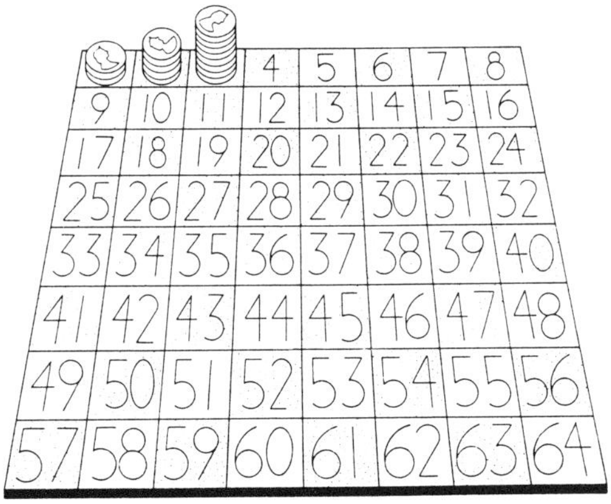

million kilometers from Earth), and in fact would reach the nearest star, Proxima Centauri. And that is only for 2^{64}. To reach Slowinski's new prime you have to double up the pile of coins a further 86,179 times. You would have left the entire universe long before you got there.

Why should anyone be interested in such huge numbers? There are various answers to such a question. To the mathematician, the way the prime numbers are distributed amongst all the numbers is an extremely interesting question in its own right. No one can say just where the next prime number will be. With small numbers, there appear to be lots of primes about, for instance of the numbers less than 25 the numbers 2, 3, 5, 7, 11, 13, 17, 19, 23 are all primes. But as soon as you start looking at much larger numbers, the primes become much less frequent, though they do not appear to follow any particular pattern.

Besides this perhaps esoteric interest, like almost all pure research there are various useful offshoots from the work. For instance simply to get the computer to handle a number with 86,243 binary digits, like Slowinski's, an entire discipline of computer science known as multi-precision arithmetic has had to be developed, and you can bet your last prime that the NSA (amongst others) is interested in that.

FIGURE 1. Exponential heights. Place two coins on the first square of a regular checkerboard, four on the second, eight on the third, and so on, doubling the number of coins each time. How high is the pile you put on the final, 64th square?

Postscript. For subsequent discoveries of large primes, see also Chapters 2, 4, 8, 63, 105, and 130.

2

How the prime number has come up

August 18, 1983

Certain subjects in mathematics are like distant relatives: you hear nothing from them for years on end and then suddenly they crop up all the time. At the moment it's prime numbers. Within a few months of the discovery of a new largest known prime number [Chapter 1] comes the publication of a new method to test whether a given number is prime or not.

There seems to be no problem to the layman. A prime number is a whole number which is not divisible (with no remainder) by any whole number other than 1 and itself, so to test if a given number is prime or not you simply look at all smaller numbers (besides 1) and check if any of them divide the number. If none of them do, the number is prime, otherwise it is composite. For small numbers this method is highly feasible, and anyone can quickly see that 2, 3, 5, 7, 11, 13, 17 are prime while 4, 6, 8, 9, 10, 12, 15 are not. Larger numbers can take a little longer. For example, is 221 prime? (Answer "No," it is the result of multiplying 13 by 17.) Five-figure numbers present quite a challenge to the average mind, but the above "obvious" method works well on a computer, and produces an answer practically instantaneously for any number less than, say 1,000,000. (Modern computers are capable of performing tens of millions of arithmetic operations in a single second.) But for numbers with, say, in excess of 100 digits, the "obvious" method is not feasible even

on a computer. Computers are still used in order to test the primality of such numbers, but they make use of other methods. The idea is to teach the computer some mathematics.

For some numbers, testing primality is easy, regardless of the size of the number. For example, the largest prime mentioned before is of a special kind: it is a Mersenne number.

Mersenne numbers are of the form $2^N - 1$ (i.e., 2 multiplied by itself N times, minus 1). There is a particularly efficient method for testing whether a Mersenne number is prime or not, known as the Lucas–Lehmer test. For $N = 86243$, a Cray-1 computer in the US took just over one hour to discover that the corresponding Mersenne number is prime. To do this by the "obvious" method would take longer than the lifetime of the universe, even using the Cray-1, the world's fastest computer. In the space available here, it is not possible to describe how the Lucas–Lehmer test works: it must suffice to say that it makes use of some advanced mathematical ideas.

However, the existence of efficient tests for certain special kinds of numbers does not help much when the problem is to determine whether a randomly chosen large number is prime or not. Moreover, this problem is not without relevance. It lies at the heart of what is at the moment the most secure coding technique in the world, so-called Public Key Cryptography.

Using this method, I could ask you to send me a coded message, explaining the coding method in an article in the *Guardian*. Using this description, you code the message and send it to me. Suppose that a bad guy tries to intercept and decode the message. Even though he knows how you coded the message, he is unable to decode it. Knowing the method of encoding does not help you decode.

An extra piece of information is required for this, something known only to the receiver, not the sender. This remarkable method, discovered by the mathematicians Ronald Rivest, Adi Shamir, and Leonard Adleman of MIT a few years back, uses large (of the order of 50- to 100-digit) prime numbers.

The potential receiver starts with two (or more) large prime numbers, which he keeps secret. What he sends to the sender is the product of these prime numbers. Knowledge of the product is all that is required for encoding, but decoding requires knowledge of the primes themselves. And whereas it is theoretically possible to factor any number into primes, there is no known method which works for very large numbers. The problem is the time it would take even an extremely fast computer to do the job.

The time required to factor a large number increases exponentially with the size of the number. The rapid growth of exponential functions means that factoring a 100-digit number is practically not feasible. Of course it is always possible that someone will come up with a clever new method for factoring large numbers. Indeed, it may have already been done, but since much of the work on this subject is classified (it is believed), we don't know. In the meantime, the best thing is to use primes as large as possible (though not Mersenne primes, as there are only 28 of them known) and hope for the best.

A good, useful method for testing primality should be capable of being executed on a computer in a time that increases linearly with the size of the number. That is, if the size of the number is doubled, so is the running time, and so on. By "size" is meant the number of digits in the number in this context. (So one can appreciate at once that the "obvious" primality test does not have this property: if the number of digits in a number is doubled, the time involved in testing primality increases dramatically, computer or not). Until recently, one of the best methods of testing primality in linear time was discovered by Robert Solovay and Volker Strassen in 1976. The main drawback with their method was that it did not produce a 100% certain result. Typically, if the method came up with the result "prime" for a given number, then you could be 99% certain that the number was indeed prime, but not absolutely sure. Would you trust your darkest secrets to a coding method which depends upon such a test?

Another method, due to Gary Miller, produces a 100% certain answer in a time which increases with the fourth power of the size of the number (which is feasible for a fast computer), but the method depends upon some mathematics which may itself be wrong. Obviously what was needed was a fast method which gives a 100% answer and which uses mathematics that we know to be correct.

In a recently published article in *Annals of Mathematics*, L. M. Adleman, C. Pomerance, and R. Rumely explain such a method. This method uses some of the ideas of the probabilistic Solovay–Strassen test, but is itself not probabilistic. Working from a pre-publication copy of the Adleman paper, H. Cohen and H. W. Lenstra Jr. have since managed to speed up the method, and their improvement is due to be published shortly.

The method is not strictly linear in its operating time. It increases with the size of the number raised to the power $C \log \log$ (size), i.e., a constant C times the logarithm of the logarithm of the size. However, because of the extremely slow growth of the logarithm function, the method works well on a computer. Typically, a fast computer requires about 30 seconds to check if a 100-digit number is prime and eight minutes for a 200-digit number. Finding a good method for factoring numbers is now the order of the day.

References

L. M. Adleman, C. Pomerance, and R. Rumely: "On distinguishing prime numbers from composite numbers," *Annals of Mathematics* 117, 1983, pp. 173–206.

H. Cohen and H. W. Lenstra Jr.: "Primality Testing and Jacobi Sums," *Mathematics of Computation* 42, 1984, pp. 297–330.

See also Chapters 1, 4, 8, 63, 105, and 130.

3

The mathematical solution to the unfiddled expenses
September 29, 1983

Working out claims for travelling expenses is a problem which confronts many professional people. In the case of a simple round-trip journey there is no difficulty: you calculate the total amount spent and claim that amount from your sponsor. But what if the journey involves visits to several places, with more than one sponsor to foot the bill? Or perhaps you wish to take advantage of your proximity to old friends during your trip in order to enjoy an evening out with them. Apportioning the expenses fairly then becomes quite difficult. One approach would be to reason that, if required to visit A, B, C, your visit to each is of no concern to the other two, and so you claim for a simple return journey from your home base to each. This method is known as Willingness to Pay, since it is based upon the fact that each sponsor is prepared to pay for you to travel there and back home again (otherwise they would not have invited you there in the first place).

In practice, because of the often large disparity between the cost of single return journeys and complex round trips, this method can work quite well, but this is not always the case. Moreover, faced with economy-minded sponsors, there may be some difficulty in persuading them that this approach is the best one. What you would like is some mathematics to back up your claim. Now it seems that some is available.

In a recent article, Bell Laboratories' Henry Pollack and Peter Fishburn address just such a problem as described above. You need to visit a certain number of places, each of which is to pay for your visit to them. You visit them each in turn in a single round trip. How do you apportion the cost between them in a fair manner?

The first step is to translate the problem into mathematical format. This involves, amongst other things, writing down the conditions which an apportionment scheme should satisfy in order to be called fair.

Mathematicians refer to these conditions, expressed mathematically, as "axioms." Then, by means of a mathematical analysis of the axioms, the mathematician seeks a "solution" to the system, that is, in this case, an apportionment system which fulfils the conditions expressed by the axioms. It is important that the axioms express the conditions involved, since the solution may well be quite unexpected.

For example, take the case of elections. As our own experience here in Britain showed just a few months ago, finding an election system which everyone agrees is fair is not at all easy, and is exactly the sort of problem for which a mathematical approach of the type outlined above is suited.

This was done some years ago by the Nobel prize winning economist K. Arrow. He wrote down a set of axioms which any fair election system should satisfy.

These axioms are all eminently reasonable, and are just what anyone would expect of a fair election system. Unfortunately, Arrow then went on to show that these axioms admit only one solution: that there is an unelected dictator who chooses the government! All very depressing. In other cases there may be no solution at all: or maybe many solutions, in which case the problem is to decide which is the "best" solution.

In their paper, Pollack and Fishburn start with three axioms for a fair apportionment of travel costs. First, the total amount paid by all the sponsors should be exactly equal to total travelling costs. Second, no sponsor should pay more than for a simple return trip from the home base to that place. Third, no sponsor should end up paying a "negative" amount, i.e., receiving money rather than paying it out. (All this is expressed mathematically, of course.)

They very quickly show that some very reasonable looking apportionment methods fail to satisfy these requirements. (One approach which fails is the so-called "Marginal Cost Procedure," whereby each institution pays out in proportion to the increase in cost incurred in including them in the itinerary over leaving them out.)

However, it turns out that the problem is one that admits more than one solution. After some analysis, the solution that they decide is "best" is called "Proportional Willingness to Pay." This is practically the method mentioned earlier. Each sponsor is charged according to the cost of a simple return journey from the home base to that one place, with the amounts being scaled so that the total claimed equals the total amount spent.

An "obvious" result true enough. But other equally "obvious" solutions have been ruled out by this mathematical approach, and as the election example shows, what is "obvious" is not always correct, so progress has certainly been made.

4

Great Minds

October 13, 1983

A popular figure in science fiction is the scientist of the future who is able to "plug in" his brain to a powerful computer. Such a possibility would produce an extremely powerful unit, combining the enormous computing power of a modern computer with the intuition and insight capacity of a human brain. Well, the day when it is possible to form a direct link between the brain and a computer may be far away, but already there are signs that close cooperation between the human mind and the computer is beginning to produce results. As regular readers of the *Guardian* will be aware, this past year has seen considerable activity in the study of prime numbers (whole numbers having no exact divisors other than themselves and 1, such as 3, 5, 7, 11). The latest event was the discovery, on September 19, of yet another largest known prime number, this time the giant number $2^{132049} - 1$.

Most of the media attention concerning such discoveries focuses upon the computer power required to find the new prime. The two previous record primes were discovered using a Cray-1 computer, for many years the most powerful computer in the world. The latest discovery was made using the new Cray XMP, which should hold the world power record until the arrival of the Cray-2 in the summer of next year.

With all the talk of computing speeds and megadollar machines, it is easy to overlook what will probably turn out to be the most significant aspect of the work. Making use of very efficient and highly sophisticated mathematical methods for primality testing, a powerful modern computer can check the primality of a particular number in a reasonable length of time.

But large primes are quite rare ("most" large numbers do have factors), so how does one go about finding the new record prime? In fact, on the face of it the task would appear impossible. Mathematical reasons dictate that a record prime has to be of the form $2^N - 1$ for some number N (i.e., 2 multiplied by itself N times, minus 1).

Prime numbers of this type are exceedingly rare indeed. Including the latest discovery, only 29 are known amongst the infinitude of all numbers! So the most surprising aspect of the latest discovery is that it came so soon after the previous one. Closer investigation reveals that the computer did not work alone. Human intuition played a large role in the matter.

Taking around one hour for each number (at a commercial rate of, say £5,000 per hour), it would be hopeless to get the computer to check every number of the form $2^N - 1$, one at a time. This is not what was done. By means of some highly intuitive and non-rigorous mathematics (the kind of thing that only the very best mathematicians can do with any hope of success), mathematicians had shown that if one were to draw a graph of the function $\log(2^N - 1)$ against N for primes $2^N - 1$, the result "ought" to be a straight line with gradient about 0.56 (log means logarithm to the base 2 here, not the logarithm found in books of mathematical tables.)

Using this "fact," it is possible to predict just where the "next" largest prime should be. In fact, in an article written long before the latest discovery of $2^{132049} - 1$, but only now appearing in the magazine *The Mathematical Intelligencer* (Vol. 5, No. 13, p. 31), Manfred Schroeder predicted that the (then) next largest prime would be around $2^{130000} - 1$, an astonishingly good prediction. No doubt this prediction was known to David Slowinski, the man who programmed the Cray XMP to find the new number. And of course the same method can be used to guess where to look for the next one.

As you may well imagine, finding record primes is, in itself, little more than an enjoyable pastime (though it does provide a good method for testing computer hardware and software, as well as obtaining a certain amount of publicity for the computer manufacturer). But this gradual coming together of the human mind and the computer may well turn out to be one of the most significant developments in the history of mankind.

See also Chapters 1, 2, 8, 63, 105, and 130.

5

The kilderkin approach through a silicon gate

October 20, 1983

What is the difference between a modern electronic computer and a 13th century English wine merchant? The answer is "Not as great as you might think." The major clue lies in the system of measurement used in the wine and brewing trade in England from the 13th century onwards, parts of which are still in use: 2 gills = 1 chopin, 2 chopins = 1 pint, 2 pints = 1 quart, 2 quarts = 1 pottle, 2 pottles = 1 gallon, 2 gallons = 1 peck, 2 pecks = 1 demibushel, 2 demibushels = 1 bushel or firkin, 2 firkins = 1 kilderkin, 2 kilderkins = 1 barrel, 2 barrels = 1 hogshead, 2 hogsheads = 1 pipe, 2 pipes = 1 tun.

As you can see, 13th century wine merchants in England measured their wares using a system of counting based on the number 2, what we now call the *binary system* of arithmetic. Leaving aside the wonderfully evocative vocabulary of the system, this means that they performed their arithmetic in the same way that a modern computer does.

We are so used to computers nowadays that it seems obvious that arithmetic should be performed in a binary fashion, this being the most natural form for a computer, which is ultimately, a "two-state" machine (the current in a circuit may be either on or off, an electrical "gate" may be either open or closed, etc). But this was not always the case.

When the first American high-speed (as they were then called) electric computers were developed in the early 1940s, they used decimal arithmetic, just as their inventors. But in 1946, the mathematician John von Neumann (essentially the inventor of the "stored program" computer we use today) suggested that it would be better to use the binary system of arithmetic, since which time binary computers have been the norm. (Not that this was the first time that calculating machines made use of the binary system. Some French machines using binary arithmetic were developed during the early 1930s, as did some early electric computers designed in the United States by John Atanasoff and by George Stibitz, and in Germany by Konrad Zuse.)

There is, of course, nothing special about the decimal number system we use every day. Certainly it was convenient in the days when people performed calculations using their fingers. Assuming a full complement of same, it is essential that there is a "carry" when we get to ten. The number at which a "carry" occurs in any number system is called the "base" of that system.

In base 10 arithmetic (decimal arithmetic), 10 entries in the units column are replaced by 1 entry in the 100s column, and so on. This means that we require ten "digits" in order to represent numbers 0, 1, 2, 3, 4, 5, 6, 7, 8, 9, all other numbers being composed of a string (or "word" if you like) made up from these digits.

Computers (and electronic calculators) use the binary system to perform their arithmetic. Here there are only two digits (known as "bits" short for "binary digits"), 0 and 1. In binary arithmetic there is a "carry" whenever a multiple of 2 occurs. So, counting from one to ten in binary looks like this: 1, 10, 11, 100, 101, 110, 111, 1000, 1001, 1010.

Arithmetic in binary (addition, multiplication, etc) is performed just as in the decimal arithmetic we learn in primary school, except that we "carry" multiples of 2 into the next column rather than multiples of 10. (So instead of having a units column, a tens column, a hundreds column, and so on, we have a units column, a twos column, a fours column, an eights column, a sixteens column, and so on.)

The fact that in binary notation all numbers can be expressed using just two digits, 0 and 1, makes the binary system particularly suited to electronic computers, since the ultimate construction element is the "gate," an electrical switch that is either on or off (1 or 0).

Of course, we do not use binary notation when we communicate with a computer or a calculator. We feed numbers into the machine in the usual decimal form, and the answer comes out in this form as well. But the computer/calculator converts the number into binary form before commencing any arithmetic and converts back into decimal form to give us the answer. All that is involved is a matter of notation (or language, if you like). The actual numbers are the same. 111 in binary means the same as 7 in decimal notation, just as *das Auto* in German means the same as *the car* in English.

All of this is a good excuse for bringing in the following teaser, one which can be used to demonstrate the absurdity of many of the questions beloved by testers of IQ in children. Fill in the next two members of the following sequence: 10, 11, 12, 13, 14, 15, 16, 17, 20, 22, 24, 31, 100, ___, ___.

I'll give the answer next time, in case you are not able to see it for yourself [Chapter 6].

6

Pi in the sky, or how the digits keep multiplying

November 3, 1983

Earlier this year, two Japanese mathematicians, Toshiaki Tamura and Yasumasa Kanada, used a computer to calculate the number π ("pi") to eight million decimal places. So let's look at some previous attempts at calculating π, and try to understand why there is an outside chance that their achievement is not totally useless.

If you want to calculate the circumference of a circle, you multiply the diameter by the magic number π. School textbooks usually give the value of π to be either 22/7 or use 3.14159. Neither of these is completely accurate. In fact, mathematicians have known for many years that it is impossible to give the value of π with total accuracy: expressed as a decimal, π would need an infinite number of decimal places. (Mathematicians would say that the decimal "expansion" of π, that is the expression for π written out in the usual decimal form of 3.14159... , is infinite. Those three dots are the mathematician's way of indicating that the decimal expansion continues indefinitely.)

So if it is not possible to say exactly what π is, how is it defined in the first place? Fortunately for most mathematics teachers, few math students ever ask that question. The answer is, π is defined to be the ratio of the circumference of a circle to its diameter. Which makes the above quoted formula for the circumference somewhat meaningless, of course, as well as raising a host of other questions, such as why is it the case that when you divide the circumference of a circle by the diameter, you obtain the same answer, namely π, regardless of the size of the circle.

There are various methods known for calculating π to however many number of decimal places you like, besides the rather inaccurate method of measuring diameters and circumferences of circles. Using these methods, mathematicians have, over the centuries been able to calculate π to ever increasing degrees of accuracy. By the end of the nineteenth century, the British mathematician William Shanks had spent 20 years of his life calculating π to 707 decimal place, publishing his result in the *Proceedings of the Royal Society* in 1873–74. (In 1945, a mistake was found in the 527th place, but by then Shanks was long past caring.)

In more recent times, computers have been used to calculate π. In 1962 it was calculated to 100,000 places. Prior to the more recent Japanese calculation, the "world record" was held by Dr. Kazunori Miyoshi of the University of Tsukuba, Japan. In 1981, after 137 hours of computation on a FACOM M200 computer, Miyoshi obtained π to two million decimal places, a result which required some 800 pages to print up.

Computers have increased in power so much over the past few years that it took a mere seven hours on a HITAC M-280H computer for Tamura and Kanada to obtain π to more than eight million places. In their calculation, they made use of a very efficient method for calculating π developed by the American mathematician Eugene Salamin. The two new record holders have announced that they intend to continue up to 16 million places this year.

Why bother? In all probability, calculation of π to enormous numbers of places will turn out to be little different (and a lot less interesting for all concerned) than climbing Mount Everest. But there is just an outside chance that some mathematical discovery will come out of all this work.

Because the decimal expansion of π can be calculated from a formula, the digits in the expansion do not constitute what is known as a random sequence of digits. Nevertheless, no one knows of any pattern in the expansion, and to all intents and purposes the expansion does look like a random sequence. Knowing the expansion up to any point does not seem to help you to determine the next digit. Pretty well all that is known is that the sequence does not start to repeat itself from any point on. By calculating π to a large number of places, it is just possible that some sort of pattern could be discerned. If that were the case, who knows what mathematical discoveries could follow.

Of course, knowledge of π to many decimal places provides mankind with one rather dubious benefit. People can spend their time trying to remember the expansion to a record length. According to the *Guinness Book of Records*, a certain Rajan Srinivasen Mahadevan of India is the current world champion, having memorized π to 31,811 places in 1981; the British record is a mere 20,013 places, established by Creighton Carvello of Redcar (1980). Apparently the speed at which these numbers were recited is also of some relevance (3 hr 49 min and 9 hr 10 min respectively).

In my last article [Chapter 5], I posed the problem of finding the next two numbers in the sequence 10, 11, 12, 13, 14, 15, 16, 17, 20, 22, 24, 31, 100, ___, ___. The clue was that the article was about binary arithmetic and number bases. The sequence quoted shows the number 16 expressed in bases 16, 15, 14, and so on, down to base 4, in which 16 becomes 100 (i.e., 1 in the 16s column and zero in the 4s column and units column). So the next two members of the sequence are 16 in bases 3 and 2, namely 121 and 10000. Simple, isn't it?

Postscript.   See Chapters 14, 77, 129, and 131 for updates on the computation of π.

7

An equation completed

November 17, 1983

The mathematical world was stunned earlier this year by the news that a young West German mathematician called Gerd Faltings had solved a major open problem dating back to 1922. What Faltings did was prove the Mordell Conjecture. It is, unfortunately, pretty nigh impossible to say here just what the Mordell Conjecture is, since some considerable mathematical knowledge is necessary even to state it correctly. But what can easily be explained is how Faltings' result goes part way to solving one of the most famous open problems of mathematics: Fermat's Last Theorem.

The 17th century French mathematician Pierre de Fermat can fairly be described as the father of modern number theory, that branch of mathematics that deals with the properties of the positive whole numbers. The amount of work produced by Fermat is truly prodigious, which is the more surprising in view of the fact that he was only an amateur mathematician; he was a jurist by profession.

He rarely published any of his work, which was carried out entirely in the form of private correspondence with some of the leading mathematicians of his day. Very often he would simply state his discoveries, leaving it to others to deduce how he had reached them. Many of these "challenges" remained unsolved until long after his death. Only one resisted all attempts at a solution, and has

done to this day. It is now known either as Fermat's Last Theorem or, if you think that on this one occasion Fermat was wrong and did not have a proof of his claim, the Fermat Conjecture.

The conjecture is ridiculously easy to state. Pythagoras' Theorem tells us that if a, b, c are the lengths of the sides of a right-angled triangle, with c being the longest side, then the three values a, b, c are related by the equation

$$a^2 + b^2 = c^2,$$

where x^n denotes the result of multiplying x by itself n times. This equation has many solutions. Some of them are whole numbers. For example, $3^2 + 4^2 = 5^2$, so we know that we can draw a right-angled triangle with sides 3, 4, 5 inches. The triple 5, 12, 13 provides another example. To a mathematician, this fact raises an obvious question. Is it possible to find whole numbers a, b, c such that

$$a^n + b^n = c^n$$

for any number n greater than 2? Fermat claimed that this was not possible, but gave no indication as to his reasons for making this claim.

The problem posed by Fermat (i.e., prove or disprove his claim) is, to be sure, a pleasant piece of number-theoretic fun, and had someone been able to resolve the matter soon after

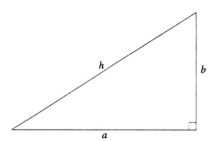

FIGURE 2.   Pythagoras' Theorem. $h^2 = a^2 + b^2$.

Fermat's death the result would no doubt be worth the odd footnote in textbooks on number theory. But no one was able to resolve it, a fact which soon led to its fame. After all, what a great story: a 17th century amateur mathematician "solves" (if so he had) a problem which resists the finest minds of 300 years' worth of professional mathematicians.

And make no mistake about it, the Fermat Problem has occupied the minds of some the greatest mathematicians since Fermat's time, and work on the problem has led to some extremely deep and far-reaching discoveries in mathematics, not only in number theory itself.

Some progress has been made on the problem. First, it is quite easy to show that if the Fermat Conjecture can be proved to be true for all prime numbers n, then it is automatically true for all whole numbers n (greater than 2).

The most significant advance on the problem was due to the 19th century German mathematician Ernst Kummer. He proved that the Fermat conjecture is true for all primes which satisfy a condition known today as "regularity." (Of primes less than 100, only 37, 59, and 67 fail to be "regular," so this settles the conjecture for all other primes less than 100.) Subsequent work has shown that the conjecture is true for a significantly larger class of primes than the regular ones. Using modern computers, it has been shown that all primes up to 125,000 are in this larger class, so at the present time the conjecture is known to

be true for all primes (and hence for all whole numbers) up to this limit.

Which brings me to the recent result mentioned at the outset. If you take the Fermat equation $a^n + b^n = c^n$ and divide through by c^n you get the new equation

$$x^n + y^n = 1,$$

where x stands for a/c and y stands for b/c. Numbers of the form of x and y, which are the result of dividing one whole number by another, are called *rational numbers*. What I have just shown is that an alternative way of stating the Fermat conjecture is that for a greater than 2 equation $x^n + y^n = 1$ has no nonzero rational solutions (though it has an infinite number of nonzero solutions which are not rational).

In 1922, the mathematician Lewis Mordell formulated a conjecture which implies that such an equation does not have an infinite number of nonzero rational solutions. In itself this is somewhat weaker than the Fermat conjecture, of course, but the conjecture actually made by Mordell applies to a much wider class of equations than the Fermat ones. (The difficulty in explaining Mordell's conjecture in simple terms lies in describing this larger class of equations).

Mordell's conjecture resisted all attempts at solution over the 50 or so years since it was first formulated, until, earlier this year, 29-year-old Gerd Faltings of Wuppertal University, West Germany, finally saw how to prove it is true, a result that has already been called the mathematical discovery of the decade.

As a result of Faltings' work we now know that the modified Fermat equation $x^n + y^n = 1$ does not have infinitely many nonzero rational solutions. Whether or not this leads to the conclusion that there are in fact no such solutions, as Fermat claimed, remains to be seen, but to go from the infinite to the finite is a significant step from a mathematician's point of view. The only drawback with Falting's proof is that it is somewhat indirect, and gives no indication as to what the maximum number of solutions may be. It could be beyond the reach of the most powerful computer. Then again it could be 10. Or even 0, which would settle the Fermat problem once and for all.

8

Winning glory by numbers

November 24, 1983

The widespread availability of fairly good computers has put within the grasp of the enthusiastic amateur the possibility of making a genuine mathematical discovery, particularly in that area of mathematics known as number theory. Two of the first people to realize this were the (then) 15-year-old high school students Laura Nickel and Curt Noll of Hayward, California. In 1975 they read about the discovery, four years previously, of what was then the largest known prime number in the world: $2^{19937} - 1$, i.e., 2 multiplied by itself 19937 times, minus 1.

Undeterred by the fact that this discovery was made by Bryant Tuckerman, working for the computer giant IBM, Nickel and Noll decided they would try to beat his record. It took them three years, but in 1978, after some 350 hours of computer time at the computer center of the California State University at Hayward, they finally managed it: they discovered the new prime number $2^{21701} - 1$.

(A prime number is any positive whole number which cannot be exactly divided by any number other than itself and 1. But even using a powerful computer, for very large numbers it is not feasible to check if a number is prime by looking to see if anything divides it: the computations would take millions of years. So other methods must be used relying upon established mathematical facts. For numbers of the form $2^N - 1$ there is a par-

ticularly efficient method to check primality, called the Lucas–Lehmer test. Consequently, record primes are always of this type. Tuckerman's prime requires 6002 digits to write out in the usual fashion; Nickel and Noll's needs 6533.)

So, in October 1978, Nickel and Noll, then only 18 years old, found that they had become instant celebrities. Their discovery was reported in every major American newspaper, and was announced on nationwide television by Walter Cronkite.

But already the writing was on the wall. As far as the search for record primes was concerned, the era of the big leaguers was about to dawn. In early February, 1979, Noll (minus Nickel this time) found another record prime, $2^{23209} - 1$. It took him nearly nine hours of computer time simply to check that this number was prime, let alone to find it in the first place.

Two weeks later, David Slowinski, a young programmer working for Cray Research at Chippewa Falls, Wisconsin, brought the immensely powerful Cray-1 computer into the hunt. It took this incredible machine a mere seven minutes to run the primality check on Noll's new number. By Sunday April 8, Slowinski's Cray-1 had established its supremacy and got the world record: $2^{44497} - 1$. Since then, the Slowinski–Cray-1 team has held the record continuously, taking it as far as the 39,751-digit monster $2^{132049} - 1$, on September 19, this year. [See also Chapters 1, 2, 4, 63, 105, and 130.]

So the amateur had best look elsewhere if a record is wanted. This is just what Paul Pritchard did early this year. An old, unsolved problem is to discover whether or not there is a maximum length for an arithmetic progression of prime numbers. An "arithmetic progression" is a sequence of numbers which increases by equal steps. For instance, 1, 3, 5, 7, 9, 11, which increases in steps of 2, or 5, 10, 15, 20, 25 (steps of 5). What the above question asks for is arithmetic progressions consisting entirely of prime numbers.

Until recently, the longest such sequence consisted of just 17 prime numbers, discovered in 1977. Then Pritchard found one with 18 numbers. The mathematics required in order to program a computer to look for such sequences is not difficult. Nor does it require a particularly powerful computer. What you need is lots of time on the machine. Pritchard, of Cornell University, USA, obtained his computer time in a particularly efficient manner.

Most university computers consist of a single, central computing unit to which are attached perhaps 20 remote terminals across the campus, where different users can have access to the machine. A typical user will type in instructions to the computer at a rate of, say, two symbols a second. Since a modern computer is capable of performing up to a million operations per second, this means that for "most" of the time the computer is sitting idle, waiting for someone to type in the next instruction.

It was this waiting time that Pritchard used in order to run his program. He instructed the computer to work on his problem whenever it had nothing better to do, and put it aside when something else came up.

With this approach it turned out that on average the computer was spending a total of 10 hours a day working on his problem. Within a month of starting, Pritchard had his world record: an arithmetic progression of prime numbers of length 18. His progression begins with the prime 107,928,278,317 and goes up in steps of 9,922,782,870 until it reaches 276,615,587,107.

9

How to put the world back in its right place

December 1, 1983

Today sees the long awaited publication, by Christian Aid in conjunction with CAFOD, of the English language edition of the Peters World Map, which Christian Aid is calling *The North and South Map*. Hitherto, the only versions of this startling new map available in this country other than the original German edition, produced several years ago, were ones produced by *New Internationalist* (the Third World magazine) for their readers, and by UNICEF as part of a mailing list publicity drive.

It is a mathematical fact that it is impossible to draw a map on a flat surface which accurately depicts the spherical surface of the globe. You have to make a decision as to what features are to be sacrificed. Geometric considerations influence the decision, but in the end human cartographers must make the crucial choices. One choice proved to have a profound influence of our perception of the world. The 16th century Flemish cartographer Gerhard Kremer's (Latin name Gerhardus Mercator) projection is without doubt the one most familiar to us today. It is the world map in the traditional school atlas.

Mercator's map has several pleasing features. It is rectangular and, at any point on the map, north lies vertically above the point, south vertically below, east horizontally to the right, west to the left. Climatic regions are readily identifiable as horizontal bands across

the map. The left-hand edge of the map is a direct continuation of the right, which means that it is possible to cut the map vertically into two pieces and rejoin the two halves along the old edges to form a new map on which a different part of the world lies centrally.

Another advantage is the one which led Mercator to devise his map. Drawn on the map, any "loxodrome" (or "rhumb line") turns out to be a straight line. A loxodrome is a line "drawn" on the Earth's surface which makes a constant angle with all lines of latitude and longitude. Consequently, a loxodrome is the path which a ship or airplane would follow if a constant compass bearing were kept. Having such navigation lines represented as straight lines on a map was obviously useful.

To obtain the Mercator projection, you start with the globe. Imagine piercing a hole in each pole and then stretching out the two ends of the globe to form a cylinder. Now cut this cylinder along its length at the 180 degree meridian and unroll it to form a rectangular map. This map will be a good representation of the world near the Equator, but nearer the poles the initial "pulling out" will cause gross distortion of shape. To eliminate this distortion, the map is stretched vertically, the nearer the poles the greater the stretching required, so that at any latitude, the amount of vertical stretching exactly balances the horizontal stretching. (None of these distortions can be performed physically; the various "stretchings" are mathematical ones which enable a map to be drawn.) The final map will then have the loxodrome property and will also represent the shape of relatively small areas of countries fairly accurately. What is sacrificed is any semblance of reality as far as the size of countries is concerned. For example, on the Mercator map, Greenland, with a surface area of around 2.1 million square kilometers, appears to be the same size as Africa whose area is some 30 million square kilometers!

The gross area distortion in Mercator's map can be somewhat avoided by drawing maps which try to preserve the spherical shape of the Earth, such as the Aitoff projection, but these maps do not have any of the advantageous properties of rectangular maps.

It is mathematically possible to construct a rectangular map of the world which does represent areas faithfully. One simple way is by the "cylindrical projection." If a rectangular piece of paper (of suitable size) were wrapped around a globe to form a cylinder touching the globe along the Equator, and all details from the globe were transferred onto the paper horizontally, and the paper unwrapped, the resulting map would reflect the relative areas of the countries of the world accurately. But there is tremendous distortion of shape away from the Equator.

In 1974, the German historian and cartographer Arno Peters claimed there was a better way. Peters' point was that familiarity with the Mercator map of the world (or relatively minor improvements to it) results in the advanced countries of Europe and North America having a totally false impression of the size of the Third World (and hence of its problems). Having thus put mathematics (in the form of cartography) on the spot, he developed the mathematics necessary to construct a new map of the world, the Peters Projection.

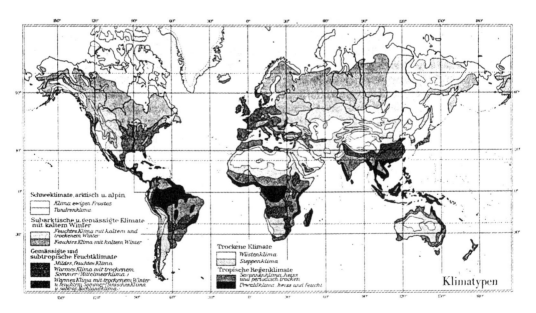

FIGURE 3.   The world. A fairly typical map projection, seen in many atlases. Good for directions, but hopeless at indicating land mass.

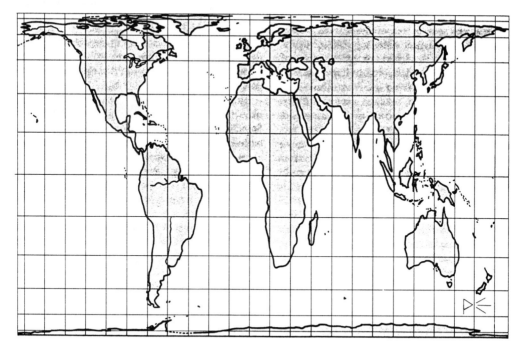

FIGURE 4. The Peters Map. The world according to Arno Peters. A faithful representation of land mass.

The basic idea behind the construction of a map is the drawing of the grid, the representation of the lines of latitude and longitude on the globe. Peters begins by drawing the grid rectangles on either side of the Equator. All other grid rectangles will have the same width as these "base" rectangles. But, because the lines of longitude on the globe get closer together as you approach the poles, in order to represent areas accurately on the Peters map, the heights of the grid rectangles must be made to decrease as you go away from the Equator.

The precise amount of decrease in height from one grid rectangle to the next required in order to keep the areas of these rectangles truly representative of the region of the globe they denote, is relatively easy to calculate, and the drawing of the grid can be left to a computer. The countries of the world can then be drawn in rectangle by rectangle. The result is a map of the Earth which accurately reflects the size of the various countries.

This projection causes some distortion of shape. The height of the Equatorial "base" rectangles is chosen so that the amount of distortion is minimal around the 45 degree latitudes, where the population density is highest. Consequently, the Equatorial regions appear somewhat elongated on the Peters map, the Polar regions shortened. But the areas are in the correct proportion. (So, in particular, the Soviet Union really is smaller than Africa!)

Of course, since the drawing of maps involves such significant choices of what to depict accurately, the appearance of any new map is bound to arouse some controversy, and the Peters Map is no exception. For instance, there are other maps which represent areas "fairly accurately," such as the Winkel Tripel projection used in Heinemann's State of the

World Atlas. But it does seem a little surprising that no English language edition of Peters' map was available until this year. The first version was prepared by *New Internationalist,* who sent it out to their subscribers in place of the June edition of their magazine.

Then in September, UNICEF used the same artwork to produce a slightly smaller edition of the map, and sent out 100,000 copies (using a UK mailing list) as part of a publicity campaign. Today's version from Christian Aid is a large wall map, intended primarily for educational purposes.

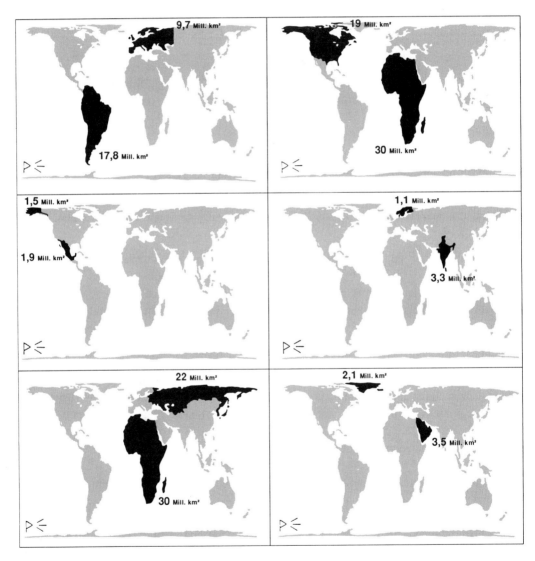

FIGURE 5.   The size of Africa. The Peters Map shows just how big are Africa and South America, compared to North America and Europe.

10

How you could take on Euler

December 8, 1983

In my November 24 article [Chapter 8], I wrote about how two American teenagers made use of the now widespread availability of powerful computers in order to make a genuine mathematical discovery, in their case, a new largest known prime number. In 1978, 18-year-olds Laura Nickel and Curt Noll discovered the new prime number $2^{21701} - 1$, thereby breaking the five-year-old record of IBM's Bryant Tuckerman.

Their discovery was made using a CDC Cyber-174 computer at the University of California; so although Nickel and Noll were really only young amateurs, they did have access to more computing power than is available to the owner of a home micro, which ought to deter micro owners from trying to better their feat. (In fact, over the last two or three years the search for record primes has been dominated by the giant "supercomputers" such as the Cray-1 and the Cyber-205.) Fortunately, there are many other elementary questions in mathematics which the home micro owner can attempt with at least a modicum of hope for success, particularly questions involving prime numbers.

A prime number is any whole number which is not divisible (without leaving a remainder) by any number other than itself and 1. Every whole number can be decomposed into a product of prime numbers, so the primes numbers constitute the building

blocks from which all whole numbers are constructed. Consequently the study of prime numbers is both important and widespread. It is also notoriously difficult. Many easily posed questions have remained unsolved over many hundreds of years, often in spite of great efforts to solve them.

For instance, in a letter written to his colleague Leonhard Euler in 1742, the mathematician Christian Goldbach conjectured that every even number is the sum of two numbers that are either prime or else 1. It is easily seen that this is true for the first few even numbers: $2 = 1 + 1$, $4 = 2 + 2$, $6 = 3 + 3$, $8 = 5 + 3$, $12 = 7 + 5$. And computers have been used to verify Goldbach's Conjecture for all even numbers up to 100 million. (This figure was reached a few years ago by my colleague at Lancaster University, William Light. It may have been increased in the meantime). However, the conjecture has never been completely proved.

As with the search for record primes, Goldbach's Conjecture is probably not a wise choice for the amateur armed with only a micro-computer, though it does illustrate quite well how simple questions can turn out to be extremely hard to answer. In my November 24 article [Chapter 8], I did describe one problem about primes that an amateur could tackle. Another one worth having a look at concerns formulas which generate primes.

It is not widely known, even amongst mathematicians, that there are simple formulas which generate all the prime numbers (and no other numbers). One such is the following. (If you are not used to handling mathematical formulas, you will probably not agree with my describing this one as "simple," but don't worry. It is included here for the benefit of the dedicated computer buffs only.) Given whole numbers (positive) M and N, first calculate the number

$$K = M(N + 1) - (N! + 1).$$

($N!$ is the mathematicians' shorthand notation for the product

$$N \cdot (N - 1) \cdot (N - 2) \cdot (N - 3) \cdots 3 \cdot 2 \cdot 1.$$

So, for example, $3! = 3 \cdot 2 \cdot 1 = 6$, $4! = 4 \cdot 3 \cdot 2 \cdot 1 = 24$, $5! = 120$, $6! = 720$.) Then calculate the number

$$P = \frac{1}{2}(N - 1)[\text{ABS}(K^2 - 1) - (K^2 - 1)] + 2.$$

($\text{ABS}(X)$ denotes the "absolute value" of any number X, that is the magnitude of X ignoring any minus signs. So, for example, $\text{ABS}(5) = 5$ and $\text{ABS}(-5) = 5$.)

For any starting values of M and N, the answer, P, you get by the above formula will be a prime number, and all prime numbers will appear as a value of P for some value of M and N. The prime number 2 (a very "odd" prime, because it is not odd as are all the rest) is a value of P for infinitely many values of M and N; each other prime number is a value of P for just one choice of M and N. A nice exercise in computer programming is to write a program that calculates P as above. (But notice that the number $N!$ becomes very large for even modest values of N, so the program can only be run for a few values of N.) Mathematicians can also have fun trying to prove that the above formula does as claimed. This is not particularly difficult. It depends upon a standard mathematical result known as Wilson's Theorem.

But though there certainly are simple (!) formulas that produce all (and only) the primes as values, it is known that there is no *polynomial* formula which does this. For instance, to take the simplest case of a polynomial formula, namely a quadratic polynomial of the form

$$P = A \cdot N^2 + B \cdot N + C,$$

where A, B, C are fixed numbers, it is known that there are no values of A, B, C such that the resulting formula gives only prime values of P for all values of N. But an interesting question is what is the longest sequence of primes that such a formula can generate when you take the successive values $N = 0, 1, 2, 3$, etc.?

In 1772, Euler noticed (Heaven only knows how or why!) that the quadratic polynomial formula

$$P = N^2 + N + 41$$

gives a prime value for P for $N = 0, 1, 2$, etc., right up to 39; that is an unbroken sequence of 40 primes. It is quite easy to write a computer program that looks for values of A, B, C for which the formula

$$P = A \cdot N^2 + B \cdot N + C$$

gives a longer sequence of primes for $N = 0, 1, 2$, etc. After all, Euler found his example long before computers were around.

Using a VAX-11 computer at Lancaster University, I searched for values of B and C right up to 10,000 that might work in the case $A = 1$. ($A = B = 1$ in Euler's formula, of course.) I could not find any. Round One to Euler. (If you allow negative values of B and C then longer sequences can be found. For example, with $B = -79$ and $C = 1601$ you get 80 primes, but this example is really just a disguised version of the Euler formula, so Euler still wins.)

The reason why I carried out the above search with a single value of A (namely $A = 1$) was to cut down the computer time used. To search through all values of B and C from 1 to 10,000 means looking at 100 million different cases. But that leaves plenty of other possibilities for owners of home micros to have a look at. In fairness, it should be said that there is some evidence that Euler's original formula cannot be bettered, but this is by no means a certainty. So if there is anyone out there with a micro which is doing nothing, you could give it a try.

Addendum published on December 22. In answer to many letters from readers let me stress that there was no misprint in the formula for generating primes quoted in my article on Euler on December 8. Try substituting in the values $M = 1$, $N = 2$. This gives $P = 3$, the first odd prime. $M = 5$, $N = 4$, gives $P = 5$, the next prime. The rest I'll leave to you.

Postscript.   See also Chapters 16 and 20.

- *Time*'s Man of the Year: Peter Ueberroth

- Baby Bells are Born

- Reagan wins reelection

- Mary Lou Retton brings home
  the gold

- Indira Gandhi assassinated

- David Lean's *A Passage
  to India*

- Bhopal (India, gas leak)

- David Mamet gets
  Pulitzer Prize for
  Glengarry Glen Ross

- LA Olympics boycotted by
  Communist Bloc

- House bans US aid to Contras

- Desmond Tutu wins Peace Prize

- *Purple Rain*

- *Miami Vice* premiered

- *Splash*

- *Beverly Hills Cop*

1984

11

The good guys sometimes win

January 5, 1984

The central theme of the film *War Games* is a computer simulation of a global thermonuclear war, in which the computer examines a number of different scenarios, to see if any of them lead to a "victory." The use of computers for such simulations is widespread. One particularly interesting example of this concerns an intriguing mathematical paradox known as the Prisoner's Dilemma, which was formulated in 1950 by Merrill Flood and Melvin Dresher.

Imagine that you and a colleague commit a crime, and that you are both picked up by the police for questioning. The police have no evidence against either of you, though they are fairly sure that you are the culprits. They separate you and question you both in isolation. They make the same offer to you both: "If you tell on your accomplice and he tries to keep quiet, you will get off for free and he will get 15 years. If you both keep quiet and don't admit anything, we shall frame you both and you'll each get a two-year sentence."

"But what if we both tell on the other?" you ask.

"Well, in that case we shall prosecute you both on the two testimonies and you will each get a ten-year sentence. After all, the public expects to see someone being punished."

What do you do? If you both keep quiet, the worst that can happen is a two-year sentence, and maybe not even that if the police botch rigging the evidence against you. On the other hand, why not tell on your mate? That seems a reasonable offer at first glance. Provided he keeps quiet, you get off free. But wait a minute. Your accomplice will also think like this, so probably what will happen in the end is that you both tell on the other, and so you both get ten years. Jointly speaking, the worst possible outcome.

So why don't you both keep mum? This is where it becomes paradoxical. You are both kept in isolation, so neither of you has any contact with the other. So what *you* decide to do cannot possibly have any influence upon what the other does. Even if you had both decided before your arrest to keep quiet, the urge to tell at the last moment and escape as a free man is irresistible. Even though you can reason that your accomplice will have the same urge, resulting in you both getting almost the worst of all options. You are logically forced into making what will inevitably be a "bad" choice. After all, you would be a fool to keep quiet. Your accomplice could then get off free, leaving you to serve 15 years.

The Prisoner's Dilemma is not solely concerned with individuals who transgress the law. On a global scale, it perhaps explains why the superpowers spend such vast amounts of money on expensive weapons systems. It would be better for both sides to spend their resources on more pressing needs at home.

But then they run the risk of the other side developing a new weapon in secret and blackmailing them into submission. So both sides spend ever more on defense, knowing that their opponents are doing precisely the same, so that neither side can ever be in a position to gain any advantage from their efforts. The effect is a continual and costly drain on national resources. This lends considerable extra interest to the computer simulation of the Prisoner's Dilemma, organized last year by Robert Axelrod of the Department of Political Science and the Institute for Public Policy Studies at the University of Michigan in America.

He formulated the problem in terms of two options, "cooperate" and "defect." The two protagonists must decide which option to take. If protagonist A decides to cooperate while protagonist B chooses to defect, then B gains a huge (numerical) advantage: the bully wins over the pacifist.

If both A and B defect, both suffer some degree of hardship (measured numerically). If both choose to cooperate, they both gain (points), though not to the same extent as the defector over the cooperator. The essence of the thing is that both A and B must choose to cooperate or defect at the same time, without knowing what the other will do. So, if A and B play this game once, it is exactly the Prisoner's Dilemma as described above. But what if, instead of just two participants, there are of the order of 50? Each of the 50 players will play the cooperate/defect choice game with each of the other participants in turn. And this entire round-robin tournament is repeated many times, with each player building up a profile of each of the opponents.

What Axelrod did was to arrange for a number of people (62, in fact, from six different countries) to work out a strategy for playing the iterative game described above, and to write their strategy in the form of a computer program. He then got all of these programs to play each other on the computer, over and over again, each one building up a record of the past behavior of the others. In such a situation, which clearly bears a close resemblance to the behavior of countries in the world, each adopting their different postures, who does better in the long run, the good guys or the bad guys?

The answer is, for most of us, encouraging. After a hundred or so iterations of the round-robin, most of the bad guys had been eliminated (lost all their points), while the moderates were getting along fine. (The really nice guys soon get trounced of course: there are limits, even in a computer game).

The most successful strategy submitted was one called *TIT for TAT*. This simply begins by trying to cooperate and thereafter does whatever the current opponent did to him the last time they met.

Given a sensible hand at the top, this would seem to provide mankind (at least the good guys) with some hope for the future—provided we survive the first hundred rounds of the game.

12

Patterns and palindromes

January 19, 1984

The number 7 has a rather peculiar property. If you work out its reciprocal as a decimal, the result is a pattern of six digits which repeats itself ad infinitum:

$$\tfrac{1}{7} = 0.142857\ 142857\ 142857\ldots$$

If you multiply this number by any of 2, 3, 4, 5, 6 (to obtain the decimal representation of 2/7, 3/7, 4/7, 5/7, 6/7, respectively), you get the same pattern shifted along. For example:

$$\tfrac{2}{7} = 0.2857\ 142857\ 142857\ldots$$

$$\tfrac{3}{7} = 0.42857\ 142857\ 142857\ldots$$

But just how special is this property of the number 7? Are there any other numbers N such that the decimal representation of $1/N$ consists of an infinite repeating pattern which is reproduced (up to a shift) when the number is multiplied by each of the numbers 2 to $N-1$? Can you find any such numbers N less than 100, other than 7 itself?

It is possible to attack this problem mathematically, but the simplest approach is to make use of your micro. Write a program which computes the reciprocal of a given number, storing one digit in each word using an array. Look for repeating patterns. (There is an easy way to do this within your routine for calculating the reciprocal. I will leave you to discover what this is.)

Each time you find a number whose reciprocal is a repeating pattern, compare it with the patterns you get by multiplying by 2, 3, etc. This will require your writing a routine for multiplying your "multi-word" decimals by 2, 3, etc. (An alternative method is to repeatedly add your reciprocal to itself the required number of times, checking for a repeat pattern each time.) Once you have your program working, you should very quickly be able to obtain a listing of each of the numbers N less than 100 with the above property.

It is probably a good idea to get a printout of the repeating pattern of each number you find. It may help you spot some sort of general rule which governs the behavior of numbers with this curious property.

Another obvious problem for a computer attack involves "palindromic" numbers, numbers such as 1221 which read the same backwards as they do in the usual, forward sense. (I assume everyone is familiar with linguistic palindromes, sentences which read the same in both directions, such as Adam's greeting to Eve in The Garden of Eden: "Madam, I'm Adam." Palindromic numbers are just the mathematician's analogue of these.) Our question concerns palindromic squares: find numbers N such that $N \times N$ is palindromic. Many examples are known, e.g.,

$$11 \times 11 = 121$$
$$26 \times 26 = 676$$
$$264 \times 264 = 69696$$

It is not hard to demonstrate that there are infinitely many such palindromic squares. And it is also quite easy to write a computer program which looks for and prints out palindromic squares. If you do this, you will soon see that practically all your results are palindromes with an odd number of digits. In fact, unless you go as far as 836, you will not see any palindromes with an even number of digits. Then comes $836 \times 836 = 698896$.

Are there any other examples of palindromic squares with an even number of digits? It may be that in order to find any you need to program your computer to handle numbers larger than the word size (so-called multiple precision routines). This is also quite straightforward, and practically a must for anyone hoping to use a micro to do some mathematical research.

The simplest way is to write routines that perform addition, multiplication, and so on with multiple word numbers in a way analogous to that we ourselves use to perform arithmetic longhand, with "carries" into the next word whenever a certain length of number is exceeded. Or you could use the "one digit per word" approach I suggested for the first problem. Though in general very wasteful of storage space in the computer, for problems such as we are considering here this is entirely feasible. (There are much more efficient methods for doing it, but for numbers of up to, say, 50 digits the "obvious" methods work well.)

Don't forget to check that your arithmetical subroutines work correctly before you set the computer going on a long search for palindromic squares! There is nothing more annoying than to find that a weekend's computing has been wasted chasing round some erroneous loop in a multiplication routine.

One final problem which lends itself to a straightforward computer solution is this. Arrange the digits 1 to 9 into two numbers so that one is the square of the other. (All digits must be used twice.) As an initial hint, let me suggest that before you write your program you think about how many digits must be in each of the two numbers. I'll give them in a few weeks' time.

Postscript. See Chapter 17 for solutions and an update.

13

Challenging the theory of safety in numbers

January 26, 1984

The mathematics underlying secret codes has been in the news again with the claim by Arnold Arnold that he has developed a method for breaking the so-called public key codes which are in widespread use by the world's military, financial, and political organizations (*Guardian,* January 12). I remain highly skeptical (professionals are habitually cautious), but in all the excitement there has been a tendency to overlook the fact that there has recently been a major and quite unexpected advance in computer mathematics which, though it has not destroyed the public key system, has certainly dented its armor.

The mathematical idea which lies behind public key coding is that it is very easy for a computer to multiply two numbers, but extremely difficult to reverse the process and find two numbers whose product is equal to a given number. Not that this difficulty is confined to computers. For instance, it would not take you long to work out that $2011 \times 877 = 1,763,647$. But if I asked you to find two numbers which, when multiplied together, give 366,893, you would have quite a task on your hands. (The answer is given at the end of this article.) Of course, a computer would be able to factor this number (i.e., split it into two numbers whose product it is) in a fraction of a second, simply by looking at all the possible combinations that there are of two numbers. But for larger numbers, say of the order of 50

digits or more, the computer is in the same position you are with numbers of more than five digits.

The idea behind the public key system is that encoding a message is related to multiplication of very large numbers whereas decoding is related to factoring large numbers. The exact correspondence is a bit technical, and in any case is not really relevant to this article.

What is relevant is that this application of elementary mathematics has led to considerable interest in the problem of factorization over the last few years, with some of the world's best mathematical brains joining in the attack. Not surprising, with so much money and expertise being thrown at the problem, something was eventually bound to give. But, as is often (always?) the case, when the breakthrough came it was almost by accident. A chance remark over a glass of beer threw the mysterious world of encryption experts into a (short-lived, as it turned out) turmoil.

On the face of it, if you are asked to factor a large number there does not seem to be any alternative to looking at all possible combinations of smaller numbers, multiplying them together and looking to see if the result is the number you started with. But in fact there are other methods, and they do not involve any great mathematical knowledge to understand. One particularly simple method was discovered by the great 17th-century French mathematician Pierre de Fermat. (See the sketch in the box following this article.)

Methods such as Fermat's greatly simplify the problem of factorization, but substantial problems remain. Until very recently, factorization of 50-digit numbers seemed to be the limit of what could be achieved, and even then it was necessary to use the world's most sophisticated computers such as the Cray-1. With the Cray-1 stumped by 70-digit numbers, such numbers were thought to be adequate for coding purposes.

Then, in the autumn of 1982, at a scientific meeting in Winnipeg, Canada, factoring experts Marvin Wunderlich and Gus Simmons went out for a beer with Tony Warnock, a computer engineer with Cray Research. The two mathematicians were discussing just what it was that made factorization so difficult: in essence that, no matter what method you use, in the end you have to look at a great many possibilities, each differing only slightly from · the others, and test each one *in turn*.

Warnock at once pointed out that, because of the way in which the Cray-1 performed its arithmetic, this process could be greatly speeded up by checking many different possibilities *at the same time*.

With hindsight, Warnock's suggestion seems obvious, as is generally the case with any new idea, but it proved to have tremendous consequences, and has largely altered the course of attempts at factorization. It is now believed that there is not much chance of mathematicians coming up with any new factoring technique in the near future, so the emphasis has switched to the actual machines used to perform the computations.

The Cray-1 computer can now factor 60-digit numbers in under three hours, and has managed to "crack" (technical term) in just over five hours a 63-digit number which had previously resisted several onslaughts. (To a mathematician, all numbers are interesting but some are more interesting than others, and it is these latter that tend to come in for sustained attack). These are results which everyone thought were totally impossible just a few months before. And if the Cray-1 can do this (and the Cray-1 is getting a bit long in the tooth now), what could be done using the next generation of computer now coming into operation?

Wunderlich is currently working with a new computer called the MPP (Massively Parallel Processor), which belongs to NASA, who uses it to analyze satellite data, and expects to be able to crack 60-digit numbers in about an hour once his program is running. Another group, at the University of Georgia, is building a special computer designed specifically for factoring, called the EPOC (Extended Precision Operand Computer), or, more colloquially, the Georgia Cracker. With a computer designed specially for factoring, who knows what might by achieved?

At the moment, the inventors of the public key system are playing safe and advising users of the system to employ 200-digit numbers in their codes. These seem to be safely out of range still.

• In answer to the question posed above, the factors of 366,893 are 193 and 1901. Micro owners can have fun emulating the big boys by programming their machine to factor numbers using Fermat's method, described in the box.

FERMAT'S FACTORIZATION METHOD

Suppose you want to factor a number n into two factors a and b. (Once you can do this, you can completely factor any number into primes by applying the same technique to a and b, then to their factors, and so on.) This is mathematically equivalent to finding two numbers x and y such that $n = x^2 - y^2$.

For, if you have x and y with this property, you get $a = x + y$ and $b = x - y$, making use of the algebraic identity $x^2 - y^2 = (x + y)(x - y)$.

Conversely, if $n = a \cdot b$, then (provided n is odd, which we may assume from the outset since even numbers are easily recognized and any factors of 2 can be factored out before you start) $x = (a + b)/2$ and $y = (a - b)/2$ will give $n = x^2 - y^2$.

The idea now is to find numbers x and y with the above property, or what is the same thing such that $x^2 - n = y^2$.

What you do then is to start with the smallest value for x such that $x^2 - n$ is positive, and then keep increasing x by 1 until $x^2 - n$ works out to be a perfect square, in which case y is its square root and the number is factored. This can be speeded up by using special tricks to see if a number is a perfect square or not. For instance, only numbers that end in 0, 1, 4, 5, 6, 9 can be perfect squares, so numbers ending in 2, 3, 7, 8 can be disregarded at once. Other, more sophisticated, improvements make this a feasible method for factoring 10-digit numbers by hand and 50-digit numbers by computer (without exploiting the structure of the computer in any way).

14

The Japanese thrive on a diet of pi and chips

February 2, 1984

On November 8 last year [Chapter 6], I wrote in this column about the calculation of the number π ("pi") to a record 8 million decimal places. Just before Christmas, I received a letter from one of the two Japanese record breakers, Yasumasa Kanada of the University of Tokyo Computer Centre, informing me that they have subsequently shattered this record by going up to 16 million decimal places.

It is a feature of the methods used to calculate π to many decimal places that there is a possibility of errors occurring "well out" along the decimal expression. To guard against this, the standard procedure is to make a second calculation of π using a different method and compare the two results. Using an S-810 Model 20 supercomputer made by Hitachi, Kanada made a second calculation of π to just over 10 million places of decimals. By comparing the two results, he concluded that he (or rather his computer) now knows the decimal representation of π accurately to 10,013,395 places. For those that are interested, the 10 millionth digit in the expansion is 7.

The computer used for this calculation, mentioned above, is one of a small group of very advanced machines which are so much more powerful than "ordinary" computers that they are referred to as "supercomputers." The Cray-1 and the Cyber-205 (both American) are examples of such machines. Comparison of the speed and power of supercomputers is very difficult to make, since any machine will have various idiosyncracies that make it more suited to one type of calculation than another. The Cray-1 has hitherto been regarded as the "world's most powerful computer" because of its very short "clock time" (the time taken to change from one discrete state to another), but for many types of calculation the Cyber-205 will perform much better.

In fact, neither the Cray-1 nor the Cyber-205 can be indisputably described as faster than the ILLIAC-IV computer, a "one-off" computer built by the Burrough's Corporation in the early 1970s for use by NASA. (ILLIAC-IV, now no longer used, was the first large-scale computer to make use of silicon chip technology for its central memory.)

One measure of the calculating speed of any computer is to count the number of single arithmetic operations that can be performed in a second. (The figure must include the time taken to collect the numbers from memory and store the result in memory.) Home micros are capable of handling a few hundred arithmetic operations a second. For supercomputers you have to talk in terms of "megaflops." One megaflop is one million arithmetic operations per second. The Cray-1 and Cyber-205 have, according to their manufacturers, peak

(!) computing speeds of around 200 and 400 megaflops, respectively.

According to Kanada, his Hitachi computer averaged over 450 megaflops for the entire 24 hour period that the calculation took. While not for a moment pretending to be an authority on supercomputers, that would look to me as though the world computer speed record has just passed from America to Japan.

Why calculate π to so many decimal places? One reason of course, is that, as with the search for record prime numbers, such feats provide some sort of measure of the power of a computer in a form that everyone can understand. But in the case of π there are other, mathematical reasons. As you probably know, π is defined to be the ratio of the circumference of any circle to its diameter (the size of the circle is irrelevant here, the ratio works out the same for all circles.)

Expressed as a decimal, π turns out to require infinitely many places of decimals (mathematicians say that π is "irrational" to

describe this fact), which starts off like this:

$$\pi = 3.14159\ 26535\ 89793\ldots$$

(Those dots are the mathematical way of saying that the decimal expression continues indefinitely).

However, no pattern is known in the digits in the π decimal, and there is a slight chance that by examining long expressions for π it may be possible to discern one. Kanada has used his computer to perform various tests on the number. He has shown that, although the digits cannot constitute a "random" sequence of numbers (because they are obtained from a formula), they do behave like one, satisfying various standard tests for "randomness." In other words, there does indeed seem to be no pattern.

Postscript.   See Chapters 77, 129, and 131 for an update on calculating π.

15

Another slice of pi

February 16, 1984

Two weeks ago I wrote about the recent calculation of the mathematical constant π to 16 million decimal places by two Japanese mathematicians. One reason why mathematicians show so much interest in π is that it keeps cropping up in the most unexpected places. For instance, if you were to take a stick 1 ft. in length and throw it down onto a surface on which there are drawn a series of parallel lines each 2 ft. apart, the probability that the stick will end up touching a line is precisely $1/\pi$.

Actually, the above example is not too surprising upon reflection. π is defined to be the ratio of the circumference of any circle to its diameter, so we expect π to be related to questions involving circles, and the probability of the above stick landing across a line is obviously connected with the angle between it and the lines, so the circular rotation of the stick as it falls is a relevant factor. Other examples are not so easily digested, at least not without some knowledge of advanced mathematics.

Infinite sums often involve π in their answer. For instance, the infinite sum

$$1 - \tfrac{1}{3} + \tfrac{1}{5} - \tfrac{1}{7} + \tfrac{1}{9} - \tfrac{1}{11} + \cdots$$

has the answer $\pi/4$. What is an "infinite sum"? Those dots in the above are meant to indicate that the sum continues indefinitely, following the pattern set up in the terms shown (in this case, increase the denominator in the fraction by 2 and keep alternating plus and minus signs). Of course, since there are an infinite number of terms to be added and subtracted, there can be no possibility of anyone actually arriving at the answer by sitting down with a pocket calculator and working it out like a normal (finite) sum.

To do this you would need an infinite amount of time. But mathematicians have methods whereby, in certain cases, they can calculate what the answer would be if there were an infinite amount of time available. Which is just as well, since many real life applications of mathematics require just such calculations of infinite sums.

Another infinite sum involving π is

$$1 + \tfrac{1}{4} + \tfrac{1}{16} + \tfrac{1}{25} + \cdots.$$

Here the terms in the sum consist of the squares of each of the numbers 1, 1/2, 1/3, 1/4, 1/5, etc. The answer to the sum in this case is $\pi^2/6$.

The number π also crops up when we work out some infinite multiplications. For instance, in 1650 the English mathematician Wallis discovered that π is equal to the infinite product

$$\left(\tfrac{2}{1}\right) \times \left(\tfrac{2}{3}\right) \times \left(\tfrac{4}{3}\right) \times \left(\tfrac{4}{5}\right) \times \left(\tfrac{6}{5}\right) \times \left(\tfrac{6}{7}\right) \times \cdots.$$

And if we multiply out the answer to

$$\left(-\tfrac{1}{2}\right) \times \left(-\tfrac{3}{2}\right) \times \left(-\tfrac{5}{2}\right) \times \left(-\tfrac{7}{2}\right) \times \cdots$$

we get $\sqrt{\pi}$.

Finally, a remarkably little known result about highest common factors. (The highest common factor of two whole numbers is the largest number which exactly divides into them both.) If you choose two whole numbers at random, the probability that they have no common factor (other than 1, which divides into everything) is $6/\pi^2$. For the benefit of the true math buffs, I shall give the proof of this curious fact.

Let P denote this unknown probability (i.e., the probability that two numbers chosen at random have no common factors). If you take any two numbers A and B, the highest common factor of A and B will be equal to a number N if, and only if, both A and B are multiples of N, and A/N and B/N have no common factor. The probability that a randomly chosen number is a multiple of N is obviously $1/N$. So the probability that A and B have highest common factor N is $1/N \times 1/N \times P$. Since any two numbers must have a highest common factor (if we include 1 as a possibility), if we add together all of the probabilities worked out above for all values of N the answer has to be 1 (i.e., certainty). So

$$\left(\tfrac{1}{1} \times \tfrac{1}{1} \times P\right) + \left(\tfrac{1}{2} \times \tfrac{1}{2} \times P\right)$$
$$+ \left(\tfrac{1}{3} \times \tfrac{1}{3} \times P\right) + \cdots = 1.$$

Taking out the common factor P this becomes

$$P \times \left[\left(\tfrac{1}{1} \times \tfrac{1}{1}\right) + \left(\tfrac{1}{2} \times \tfrac{1}{2}\right)\right.$$
$$\left. + \left(\tfrac{1}{3} \times \tfrac{1}{3}\right) + \cdots\right] = 1.$$

In other words

$$P \times \left[1 + \tfrac{1}{4} + \tfrac{1}{9} + \cdots\right] = 1.$$

But we know what the answer is for the infinite sum in the brackets here. As mentioned above, it is $\pi^2/6$. So $P = 6/\pi^2$, as I claimed.

16

Prime beef

March 1, 1984

The *Guardian* has a reputation for misprints. I suspect that this reputation has more to do with the quality of the readership than of the typesetters, but in any event the latter came in for quite a bit of flak following my article on Euler on December 8, last year [Chapter 10]. In that piece I quoted a formula which generates prime numbers. Many readers, upon investigating the behavior of that formula, came to the conclusion that it must have been misprinted. In fact there was no misprint, and thereby hangs today's tale.

Let me begin by quoting the formula again. Given positive whole numbers M and N, calculate

$$K = M \cdot (N+1) - (N! + 1),$$

and then calculate

$$P = \tfrac{1}{2}(N-1)[\text{ABS}(K^2 - 1) - (K^2 - 1)] + 2.$$

In this formula, $N!$ (read "N-factorial") is the mathematician's abbreviation for the number $N \cdot (N-1) \cdot (N-2) \cdots 3 \cdot 2 \cdot 1$. For example, $3! = 3 \cdot 2 \cdot 1 = 6$ and $4! = 4 \cdot 3 \cdot 2 \cdot 1 = 120$. And $\text{ABS}(X)$ denotes the "absolute value" of any number X, that is the value of X ignoring any minus sign. For example, $\text{ABS}(5) = 5$ and $\text{ABS}(-5) = 5$.

For any values of M and N, the value `P you obtain from the above formula will be a prime number. Moreover, every prime number will be a value of P for some values of M and N. So the formula does generate all the primes, regardless of anything to the contrary that you may have read in either the popular press or elsewhere. Make no mistake about it, the primes can be generated by a simple formula.

What led many readers astray last time was the way in which this formula generates primes. For most values of M and N you find that $P = 2$. But if you are patient, the other primes do appear. For instance, for $M = 1$, $N = 2$, you get $P = 3$; for $M = 5$, $N = 4$ you get $P = 5$; for $M = 103$, $N = 6$ you get $P = 7$. To get $P = 11$, however, you need to be very patient indeed: this does not happen until you try $M = 329{,}891$ and $N = 10$.

It is easy to write a computer program to calculate P using the above formula and check these results. But even without doing that, it seems clear that this is not an efficient way of generating primes. Most of the time you get the same output, 2, and only at rare intervals do you get any other output. (Question: where do you find $P = 13$?) The importance of the formula is that it does demonstrate that the primes are capable of being generated by a formula, albeit inefficiently.

While on the subject of efficiency of formulas and programs, from time to time in this column I have reported on efficient methods of testing a given number to see if it is prime. The "obvious" method of looking at all smaller numbers and checking that none of them divide it exactly (other than 1, which divides everything, of course), while perfectly adequate for computer use when the number concerned is not too large (say not larger than the word size of the computer), is hopelessly inefficient for very large numbers, say of 50 digits or more.

The fast primality tests being used are rather sophisticated mathematically, but they all start from the same idea, which goes back to the 17th century French mathematician (an amateur) Pierre de Fermat. He showed that if you take a number P which is prime, then the number $2^{P-1} - 1$ is exactly divisible by P. For large numbers P, it is much quicker to calculate the number $2^{P-1} - 1$ and see if P divides it than to look for factors of P. If P does not divide this number, then P cannot be prime. So this gives you a quick way to check that P is not prime. Unfortunately, you cannot conclude that if P does divide $2^{P-1} - 1$ then P necessarily is prime. It probably is, but there are some numbers which are not prime and yet you do get divisibility here.

This provides a nice exercise to keep your micro busy. Find non-prime numbers P for which P divides $2^{P-1} - 1$.

17

73 year old brain beats micros

March 15, 1984

The response I got when, on January 19 in this column [Chapter 12], I suggested three problems that micro owners might like to try on their machines, was far more than I had anticipated, and I was unable to reply to all who had written in to me with solutions. So, here is the current state of play.

The simplest of the three problems was to arrange the digits 1 to 9 into two numbers, one of which is the square of the other. A few moments thought tells you that one number should have six digits, the other three. It is then a straightforward matter to write a short computer program to find that the two possible solutions are

$$567^2 = 321,489, \qquad 854^2 = 729,316.$$

In fact you don't need a computer for this. Several readers solved the problem using a pocket calculator. One did not even need that. This reader solved the problem using only, as he put it, his "73-year-old brain." It took him just over one hour. This seems to be the average time required by calculator users as well. Few of the computer solvers told me how long it took them to solve the problem by writing a program, but I would doubt if it was much less than two hours before the program was tested and ready to run. The running time for the program seems to be anywhere from 40 seconds to 5 minutes, depending on the computer and, no doubt, the program used.

The second problem was to find perfect squares which are palindromic (i.e., read the same backwards or forwards) and which have an even number of digits. The only one I knew of was $836^2 = 698,896$.

This problem is quite a challenge for the average home micro, and you need to exercise some cunning in writing a program that stands any chance of success.

One micro user in Chelmsford did just this, and after 12 hours of computing came up with $798,644^2 = 637,832,238,736$, a palindromic square with 12 digits. In a postscript to his letter he adds that after almost 20 hours computing his computer found the palindromic square

$$1,042,151^2 = 1,086,078,706,801,$$

though with 13 digits this does not answer my question. (For some reason, palindromic squares seem to prefer to have an odd number of digits. Many examples of such can be found, e.g., $11^2 = 121$, $26^2 = 676$.)

I felt I was undermining the nation's economy when I discovered that a reader "somewhere in the South-east" tied up a large, mainframe IBM computer for 24 hours on a search for palindromic squares, coming up with the 16 digit

$$64,030,648^2 = 4,099,923,883,299,904.$$

As far as I know this is a world record, though whether or not the discoverer is able to claim this publicly I don't know. (Actually, my own view is that using computers for this sort of thing is quite justified.) At any rate, this seems to put the problem out of range of the home micro now. (Doesn't it?)

The third problem has quite a bit of mathematics behind it. It all starts with the observation that if you work out $1/7$ as a decimal you get the repeating pattern

$$1/7 = 0.142857\ 142857\ 142857 \cdots$$

which continues like this *ad infinitum.*

Moreover, if you multiply this number by any of 2, 3, 4, 5, 6 (to obtain the decimal representation of $2/7, 3/7, 4/7, 5/7, 6/7$, respectively), you obtain the same infinite repeating pattern shifted along one or more places, e.g.,

$$2/7 = 0.2857\ 142857\ 142857 \cdots$$

$$3/7 = 0.42857\ 142857 \cdots$$

I asked for examples of other positive whole numbers N (less than 100, say) which have the property that $1/N$ is an infinite repeating decimal which simply gets shifted along when you multiply by any one of $2, 3, \ldots, N-1$.

Of those readers who wrote in, most had quickly realised that N has to be a prime number and that the decimal representation of $1/N$ has to consist of a repeating pattern of length exactly $N-1$. Surprisingly, this turns out to be enough. A computer search will reveal that of the numbers less than 100, only 7, 17, 19, 23, 29, 47, 59, 61, 97 have the required property.

The mathematical theory behind the above "solution" to the problem is interesting, and involves "congruences" and "primitive roots," topics just slightly beyond the scope of this column. The solution to the 7-problem immediately suggests a more general problem. Find non-prime numbers N such that $1/N$ is an infinite repeating decimal and K/N is a shift of $1/N$ for all numbers K less than N which have no prime factors in common with N.

World record

Postscript published on April 5, 1984.

A reader of *Microguardian* has just found a new palindromic perfect square with an even number of digits. After reading my article on March 15, Mr. Graham Lyons of Romford, Essex, worked out an efficient method of searching for palindromic squares.

A number is palindromic if it reads the same in both directions, such as 707 or 14541. Many perfect squares are palindromic, for example 11 squared is 121 and 836 squared is 698,896. This last example was one of only three examples known of palindromic squares with an even number of digits, the other two being the squares of 798,644 and 64,030,648.

Using an IBM-XT microcomputer running a compiled BASIC program, Mr Lyons has now discovered a palindromic square with 22 digits. This is a new world record in a field which has seen considerable activity over many years, and is quite an achievement. The new palindrome is the first one above the 16-digit example mentioned earlier, and it took Mr Lyons' program an entire weekend's computing to discover. The number is the square of 83,163,115,486.

18

Shades of opinion
March 29, 1984

How would you like to use your micro to earn $100,000? Sounds too good to be true? The cash is being offered by the Fredkin Foundation of Boston, Massachusetts. All you have to do in order to claim your prize is write a computer program which subsequently makes a genuine mathematical discovery of acceptable importance.

Who decides whether your computer's discovery is of sufficient merit to earn the prize? There is no catch to this part: the final decision rests with a 12-member committee headed by Woodrow Bledsoe, a computer scientist at the University of Texas at Austin. The catch lies in the stipulation that, according the Bledsoe, "The prize will be awarded only for a mathematical work of distinction in which some of the pivotal ideas have been found automatically by a computer program in which they are not initially implicit."

So there you have it. The computer must somehow make part of the discovery itself, and not just be the workhorse of a clever mathematician. Despite some quite considerable uses of computers in mathematics during the past few decades, I do not think any discovery has yet been made which would come close to qualifying for the prize now being offered. What progress there has been has largely been in the form of the speed and computing power of the computer being used to perform essentially routine calculations in order to verify what mathematicians had already guessed.

In fact the situation is somewhat worse. There are very few instances where the use of a computer has played any significant role in making a genuine mathematical discovery.

Such feats as discoveries of large prime numbers I discount as not being real mathematical discoveries at all. The real discovery as far as this example is concerned was made by Euclid over 2,000 years ago: namely that the number of primes is infinite. The closest that a computer has come to making a mathematical discovery was in the solution of the famous Four Color Problem a few years ago.

The Four Color Problem (now the Four Color Theorem) concerns a question about the drawing of maps. In 1852, shortly after he completed his studies at University College, London, Francis Guthrie wrote to his brother Frederick, who was still a student at the college, pointing out that as far as he could see, every map drawn on a sheet of paper can be colored with only four colors in such a way that any two countries which share a stretch of common border are colored in differently.

Francis wondered if there were some mathematical way of proving this fact— if fact it were. Frederick passed on the problem to his professor, the famous mathematician Augustus de Morgan. Although de Morgan's great abilities were not sufficient to answer the question, he was able to make some progress on it.

For instance, he proved that it is not possible for five countries to be in a position such that each of them is adjacent to the other four, a fact which at first sight would appear to answer the original question affirmatively, but which on closer inspection turns out to do nothing of the sort. (Though, over the 124-year period between the posing of the problem and its final solution, many amateur mathematicians, rediscovering de Morgan's proof, claimed thereby to have solved the Four Color Problem.)

Two things are quite easy to establish about coloring of maps. First, there are simple maps which cannot be colored using only three colors, so at least four colors are required. (For instance, to color the map shown, four colors are needed.)

Second, five colors always suffice. Proving this latter fact was itself something of a triumph. For the difficulty of the problem lies in the fact that it asks about all possible maps, not just some particular map, such as a map of the world. So there is no hope of proving that four colors suffice by looking at any particular map, no matter how complicated it is. Even a computer cannot consider all possible maps, for there are infinitely many such.

Nevertheless, in 1976, Kenneth Appel and Wolfgang Haken did succeed in proving that four colors are always enough, and moreover their proof did involve the essential use of a computer (three computers, in fact). The proof rested upon some previous work of Alfred Bray Kempe, a London Barrister who, in 1879, produced what turned out to be a false proof of the theorem. Kempe's main idea was this: first you show that in any map which required five colors to color properly, certain special configurations of countries must occur. Then, by examining each of these special configurations in turn, you show that none of them can occur in a map which requires five colors. Taken together, these two results imply that no map requires five colors, which proves the Four Color Theorem.

Kempe's overall strategy was correct, but unfortunately it turned out that the class of

FIGURE 6.   The Four Color Problem. How many colors does it take to color a map so that no two adjacent countries have the same color? The Four Color Theorem says that four colors suffice, for any map, real or fictitious.

special configurations involved contains some 1,500 arrangements, each requiring a detailed analysis. It required some 1,200 hours of computer time to carry out this analysis, but in the end, the question was resolved. To be sure, the computer had not solved the problem on its own; a great deal of man's ingenuity went into the solution.

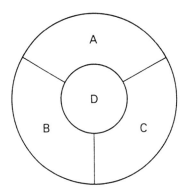

FIGURE 7.   A four-color map. This simple map requires four colors to be colored in such a way that no two adjacent countries are colored the same.

But the mathematician cannot lay sole claim to the result either: the use of the computer is essential in the proof. This fact led several prominent mathematicians to reject the new proof at the time as not really a "proof." In some sense this is true, of course. It was not a proof in the sense of this word which mathematicians had understood for centuries. But times change, and the computer has brought with it a new meaning of the word.

It is unlikely that we have seen the last of mathematical proofs that only a computer can follow. But whether we shall ever see a proof that a computer itself discovered, as the Fredkin people are asking for, is another matter. My guess is that this is a long way off, but that would not deter me from looking, and I doubt if it would deter many others as well. We may yet all be surprised. Any takers?

Postscript.   See Chapter 67 for an update on the Four Color Problem.

19

The measure of all things

April 12, 1984

It is not often that a discovery in mathematics makes front page news, but this is what occurred a few weeks ago when the *International Herald Tribune* carried a substantial front page story about the discovery that a long standing conjecture in number theory known as Mertens' Conjecture was false.

The exact statement of the conjecture is given later, but in brief what it says is that for all positive whole numbers N, the value of a certain function $M(N)$ is less than the square root of N. Computers had previously been used to show that the Mertens Conjecture was true for all values of N up to 10 billion. In any other science such overwhelming evidence would clearly be taken as conclusive. Physical scientists have to rely on much less evidence than 10 billion positive experimental results with none against. But in mathematics, though evidence of this kind is usually taken as indicating a likely truth, some nasty shocks in the past have led to a more cautious attitude towards such apparently convincing data.

This caution proved well founded in this case. For, using modern fast computers in conjunction with more traditional mathematical methods, Andrew Odlyzko of Bell Laboratories and Herman te Riele of the Amsterdam Center for Mathematics and Computer Science finally succeeded in proving that the conjecture is false.

What is the value of N for which it fails? Unfortunately, their proof does not actually produce such a number. All they show is that there has to be an N for which $M(N)$ exceeds the square root of N. (Such "non-constructive proofs," as they are known, are quite common in mathematics).

All that we can safely say is that, by virtue of the previous computer searches, such an N will have to be bigger than 10 billion. Odlyzko and te Riele themselves think that such a number would be somewhere in the region of 10 raised to the power 70, a number which exceeds the number of atoms in the known universe, and thus is well outside the reach of computers, both those of the present day and any that could conceivably be built in the entire future of mankind.

To understand what the Mertens Conjecture says, you need to know a little bit about the so-called Moebius function, $\mu(N)$. [μ is the Greek letter "mu."] This is defined like this. $\mu(1)$ is taken to be 1. (This is a special case). For any number N greater than 1, $\mu(N)$ has value 0 if N is divisible by the square of any prime number. If, on the other hand, N is not so divisible, then N has to be prime or else the product of two or more distinct prime numbers; in which case $\mu(N)$ has value 1 if N is the product of an even number of primes and value -1 if N is prime or else a product of an odd number of primes.

For example, the first few values of $\mu(N)$ are: $\mu(1) = 1$, $\mu(2) = -1$, $\mu(3) = -1$, $\mu(4) = 0, \mu(5) = -1, \mu(6) = 1$. To get a better idea of the behavior of $\mu(N)$, you could write a computer program to work out the first few hundred values. You will see that there seems to be no real pattern. But despite this seemingly wild behavior, it is an easy and pleasant evening's recreation to prove some curious facts about $\mu(N)$. For instance, if you take any number K and add together all values of $\mu(N)$ for which N divides exactly into K, then the answer is always 0.

You could check this by computer, but as I have just indicated, a computer check does not guarantee the answer, even if you go up to 10 billion or beyond. So perhaps it is best to try and prove it by more traditional mathematical methods. (It is well within the reach of a keen amateur mathematician.)

Another nice result that is not hard to prove is that if N is any number greater than 2, the result of adding together $\mu(1!)$, $\mu(2!)$, $\mu(3!)$, up to $\mu(N!)$, is always 1. ($N!$ is mathematician's shorthand for the product

$$N \cdot (N-1) \cdot (N-2) \cdots 3 \cdot 2 \cdot 1.$$

So, for example, $2! = 2$; $3! = 3 \cdot 2 \cdot 1 = 6$; $4! = 4 \cdot 3 \cdot 2 \cdot 1 = 24$.)

The function $M(N)$ which figures in Mertens' Conjecture is defined like this. For any number N, $M(N)$ is the result of adding together $\mu(1)$, $\mu(2)$, $\mu(3)$, up to $\mu(N)$. Seemingly a "nicer" function than the one in the above example using $N!$. But whereas, in the above example, the answer was always 1, for $M(N)$ there seems to be no pattern at all, save that its values appear to be all less than the square root of N. Get your micro to work out a table of values of $M(N)$, together with the square root of N, and see what I mean. (Mertens himself calculated values of $M(N)$ for N up to 10,000, using pencil and paper techniques.)

Clearly there is a moral to this tale. The advent of very powerful computers has provided the mathematician with a useful tool for collecting evidence for various conjectures, but, as the Mertens example shows, computers still fall short of making the role of the mathematician a defunct one. Personally, I find it rather gratifying to know that somewhere well beyond the reach of any computer there is a number N for which $M(N)$ exceeds the square root of N. It provides me with an element of job security!

20

Hands off

April 26, 1984

By far the greatest computer of all is the human brain. Anyone familiar with any of the areas of computer science, such as pattern recognition, language translation, robot control, or artificial intelligence, will hardly fail to realize that the computer is a poor substitute. Where computers really come into their own is in rapid calculation involving large numbers. For most of us, the human brain is a pitiful ally when confronted with even quite modest arithmetical problems, and we turn to either a calculator or a computer.

Not so a certain Wim Klein from Amsterdam. If you look in the *Guinness Book of Records* you will see that he appears as the world's "best" human computer. Now in his seventies, Klein was previously employed as a "computer" at the CERN atomic physics laboratories in Geneva. Towards the end of last year he established two new records in mental arithmetic that will be very hard to beat.

On September 30, 1983, in a packed lecture hall at the DESY atomic physics centre in Hamburg, Klein was given the number 5462 and asked to express this number as the sum of four squares in 10 different ways. He did this in 43.8 seconds. One such expression is:

$$5462 = 73^2 + 9^2 + 6^2 + 4^2.$$

Can you find the other nine? In about 40 seconds as Klein did? How long will it take your micro to perform the same feat if you include the time used to write the program?

For his second feat, also at DESY, this time on November 22, 1983, Klein calculated the 73rd root of a 505-digit number in 1 minute and 43 seconds. The number used was chosen by the audience from a collection of 42 possibilities produced by the DESY computer.

Meanwhile, back in the land of the inanimate electronic computer, I have made some progress on a problem I wrote about on this page last year. I asked if there was any quadratic polynomial of the form $A \cdot N^2 + B \cdot N + C$, where A, B, C are fixed whole numbers, such that all of its values are prime numbers, for N equal to each of 0, 1, 2, up to 40. The great 18th century mathematician Leonhard Euler noticed that for $A = 1$, $B = 1$, $C = 41$ you get prime numbers as values, for N equal to 0, 1, 2, up to 39, and as far as I know this has not been bettered in the intervening 200 years. (The numbers A, B, C are supposed to be positive, by the way, though this does not seem to make any genuine difference to the problem).

Utilizing "spare" time on one of the Lancaster University VAX supermini computers over Christmas, I made two searches. First I looked at all possible combinations of values of A, B, C from 0 up to 1000. Then, fixing A to be 1, I looked at all values of B and C from 0 up to 10,000. So in the first case I looked at some one thousand million different polynomials, in the second at a further one hundred million. The result? Euler still has the world record, as I expected he would. (There are some rather vague indications that there is more to Euler's formula than meets the eye).

What is the "spare" time I mentioned above? Well, large mainframe computers and superminis like the VAX are so fast and powerful (and expensive) that it would be wasteful to let them stand idle for any length of time. Consequently, they are usually in operation 24 hours a day, 7 days a week. To facilitate this they are programmed to operate without any human intervention. A sophisticated control program called an "operating system" performs all of the duties that a micro user must perform himself.

So the user of such a computer uses the machine like this. He accesses the computer at any one of a number of remote terminals, often a considerable distance from the main computing unit. The system allows the users to write programs in whatever language he likes, Pascal, Algol, Fortran, BASIC, Assembly Language, or any number of specialized languages for specific purposes. Having written the program, if it is a short one (to run) the computer can be used in much the same way as a micro, and the user simply sits in front of a screen and watches what happens. (One difference is that a large computer may be dealing with many different users at the same time. The operating system keeps the various programs apart, so each user has no idea what else is going on in the computer.)

For long programs such as my polynomial search, the procedure is to simply ask the operating system to run the program whenever there is nothing else to do, suspending operation when something "comes up." This way, only "spare," otherwise idle, time is used; and it is totally free, since it costs exactly the same if the computer is not actually running a user's program! So it was with a clear conscience that I sat down to my Christmas dinner knowing that somewhere deep in the bowels of Lancaster University the computer was busily whirring its way through my one billion polynomials.

21

Why runners go round the bend

May 10, 1984

This Sunday, 20,000 runners will set off on the fourth London Marathon. For those who have trained properly for the event, the first 15 or 16 miles should not be too difficult. But then (and I speak from personal experience), just after the 16 mile point, the course joins the River Thames to the north of the Isle of Dogs, whereafter it follows the path of the river almost all the way to the finish on Westminster Bridge some 10 painful miles further on.

The four-mile long loop around the Isle of Dogs itself is particularly difficult: not only are there few spectators to cheer you on, at the end of it you find yourself within a few yards of a spot passed a half hour or more previously. This week's column is intended to provide the weary runner with something to contemplate while trudging along this particular stretch. In particular, you might like to ask yourself why on earth the river follows such a meandering path.

The answer is (though on Sunday morning no runner will believe it!) that the river follows a course which involves the least amount

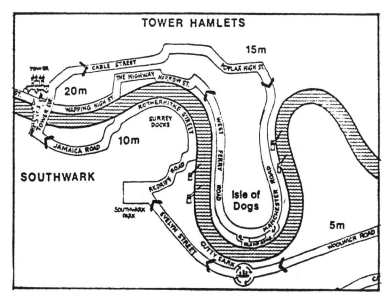

FIGURE 8. **The London Marathon.** Part of the course follows the curve of the River Thames. But why does the river itself follow such a curve? And what kind of curve is it anyway?

of work being done. This may be demonstrated with the aid of computers.

But first, why do we refer to the familiar twistings and turnings of rivers as meanders? This comes from a particular Turkish river known nowadays as the Menderes, but in ancient times as the Maiandros, a river which exhibits the twisting pattern common to all rivers.

Next question: why do rivers flow at all? Answer: gravity causes water to flow from a higher level to a lower. Supplementary question: since Nature usually acts in the most efficient manner, why are rivers not straight, for that would surely represent the most efficient way for the water to arrive at a lower level?

Your first guess at an answer is probably that, as it flows, the river encounters terrain of different types, some hard and some soft, and tends to flow along the easiest path. To anyone who has observed the planning of a motorway through the constituencies of influential MPs this seems quite reasonable; but as far as rivers are concerned it does not fit the facts.

For one thing, rivers which flow through quite homogeneous terrain also develop meanders. In particular, rivers that flow down glaciers meander, as do ocean currents like the Gulf Stream. And if you set up an artificial river flowing through a bed of sand, it too will develop meanders, no matter how straight it starts out. And—surely the clinching evidence—all river meanders are (subject to the degree of meandering) the same overall shape, wherever in the world they may be. (Take a look at a world map and see what I mean.)

So why do rivers meander? The answer, first suggested by Albert Einstein, is to be found in mathematics. Modern research has confirmed Einstein's hypothesis that a meandering river is following the most efficient path possible (i.e., the one involving the least

work) when you allow for the random motions of the water particles in the river. These random fluctuations in the passage of the water cause erosion of banks and the subsequent deposition of material to form meanders, a fairly complex process involving many variables such as the speed of the river, the type of terrain, etc. But the following simple idea should indicate the general process, and can be simulated on a computer.

Imagine a large American style city where all the streets run North-South or East-West, in straight lines a uniform distance apart. Each Sunday you set out on a 10 mile walk from your house to your favorite pub three miles away. To add interest, you stop at each road junction and choose your direction by some random procedure such as rolling a die.

Of course more often than not when you have walked your 10 miles you will not be at your pub, and must take a taxi in order to obtain your Sunday pint. But occasionally you will succeed. Suppose that each time you are successful you draw a map of your route. What kind of path will you have followed? The odds are that you will have traced out a classic meander pattern. The most likely path of a prescribed length from one point to another is a meander path.

Computers may be used to perform simulations of the kind of "walk" just described, but there is an easier way for the home micro owner to use the computer to produce "rivers." In 1951 a mathematician called Hermann von Schelling made a mathematical analysis of random walks (as tours of the above kind are called) and showed that (to a very good degree of approximation) the path of a random walk (and likewise of a meandering river) obeys the following simple rule.

Let the overall downstream direction be the x-axis. Then the angle between the x-axis and the path (or river) at any point is equal to a constant multiplied by the sine of the distance of that point from the start, *measured along the path* (not along the x-axis). It is a good exercise to write a computer program to produce examples of curves with this property, using the graphics facility on your micro. See how the constant of proportionality affects the shape of the curve. Try to produce shallow meanders and deep ones.

Of course, those of you actually running in the London Marathon will not (I assume) be carrying a computer with you at the time, and with luck your path will not be entirely governed by random motions. So as you plod around the Isle of Dogs, just console yourself with the thought that you are following the path involving the least work.

Postscript.   A number of readers wrote in to dispute the explanation of river meanders presented here. It seems that this is an issue on which there is as yet no clear consensus.

22

Further adventures in Flatland

May 31, 1984

Most people are fascinated by the idea of a fourth dimension, and many attempts have been made to try to use ideas of perspective to try to draw four-dimensional objects such as the "hypercube" a "solid" whose eight "faces" are all ordinary cubes.

A better picture emerges when you try to build three-dimensional models which are the "projections" of four-dimensional objects into three-dimensional space (just as a photograph is a two-dimensional projection of the three-dimensional object it depicts).

As a professional mathematician of many years standing, I think I have managed to develop a reasonably good mental picture not only of four dimensions but of any number of dimensions, and in common with my colleagues I can happily perform mathematics involving many dimensions of space. But this mental familiarity does not seem to help when it comes to trying to put across the idea of four or more dimensions to the uninitiated. (Incidentally, by a fourth dimension here I mean an extra dimension in space, not a mathematical representation of time such as is used in Relativity Theory.)

For this purpose, the best approach seems to be to get my listener to first imagine what life would be like in a two-dimensional world, and then see what extra things could be done by passing to a third dimension. Then, by shifting attention up one dimension, one can get some feeling at least as to what can be done in four dimensions.

In a four-dimensional world, it would be possible to mould a sheet of plastic into a "bottle" having no rim, and consequently no separate inside or outside, whose surface has only one side. In three dimensions this can only be done by allowing the surface to pass through itself. The resulting figure is known as the Klein Bottle.

The idea of a hypothetical two-dimensional world is not new. In 1884, the English clergyman Edwin Abbott published his satirical novel *Flatland,* based on just this idea, and the theme was taken further in the 1907 novel *An Episode of Flatland* by Charles Hinton.

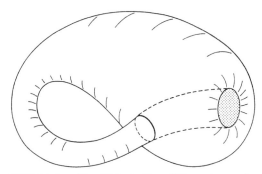

FIGURE 9. The Klein bottle. This curious object has neither inside nor outside, just one "side." In three-dimensional space, the surface of the "bottle" passes through itself; in four-dimensional space, it would be possible to construct the bottle without such self-intersection.

Then, in 1978, Alexander Dewdney, Professor of Computer Science at the University of Western Ontario in Canada, took up the idea in earnest, and began to systematically investigate the scientific laws that would govern a two-dimensional universe—the *Planiverse* in his terminology.

For humans, perhaps the most interesting question is whether there could exist intelligent life in a two-dimensional universe. It has been argued in the past that at least three dimensions are required for intelligence, since this requires a structure of brain cell interconnections far too complex to be realized in two dimensions. (In two dimensions, nerve fibers could not cross over each other, they would

have to connect.) However, computer scientists are now able to refute this objection. It is possible to construct "crossover points" in two dimensions which provide exactly the same facility as a crossover in three dimensions. All that is required is a special configuration of "computer gates," and any decent student of computer science could design and build such a device.

A fully two-dimensional computer could function just as a standard computer, the only disadvantage being that the extra circuits for the crossovers would slow down its performance. The biological equivalent of such a circuit would thus enable a two-dimensional brain to develop.

" O day and night, but this is wondrous strange "

"Fie, fie, how franticly I square my talk!"

FIGURE 10.   **Flatland.**   The original cover page from E. A. Abbott's classic tale of life in a two-dimensional universe.

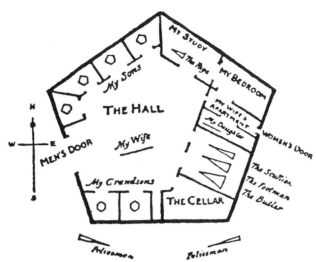

FIGURE 11.   The two-dimensional home of A. Square, the hero of Abbott's novel. Not a cross-sectional view of the house; this is the entire house!

Dewdney thought that it would make an interesting and useful exercise in computer programming to get his students to write a computer program which would simulate a two-dimensional world, using the two-dimensional television screen to realize the resulting "2D WORLD" for all to see. They postulated the existence of a two-dimensional "planet," called Arde, which to us would be just a flat circular disk. The inhabitants of this world, the Ardeans, would live on the rim of this disk, much as we live on the outer skin of our spherical Earth. Being restricted to just two dimensions would mean that Ardean life is quite different from ours (just as four-dimensional life would presumably be quite different from ours as well). For instance, when two Ardeans meet and wish to pass each other, since they cannot pass by the side (there is no "side"), one of them must climb over the other, and there would presumably be conventions for this, much the same as our Keep Left convention on the roads. All Ardean "buildings" would have to be underground, to avoid the Ardeans having to climb over them when they travel.

Travelling itself would be more difficult, since you cannot have a wheel on Arde. More precisely, you can construct a circular wheel, but you cannot put it on an axle, since this would require a third dimension. So rollers are possible, but nothing like a motor car could be built. On the other hand, airplanes could be built (using rocket propulsion) without wings: the entire airplane is built in the shape of a wing cross-section, and in two dimensions that will suffice. And so on.

Dewdney and his students developed their computer program to quite an advanced state. And then, according to Dewdney, in 1980, a rather amazing event took place. To try to explain it, Dewdney himself used the analogy of a tuning fork. If you start a tuning fork vibrating and place it close to another tuning fork of the same size, the second fork will also start to vibrate due to a resonance effect. It turned out, says Dewdney, that there really is a two-dimensional universe, with a disk-like planet populated by "intelligent beings." Dewdney's computer program created a "world" so similar to this real world that, like the tuning fork, a resonance saw set up enabling Dewdney and his students to communicate via the computer with the "Ardeans" on this planet.

In his book *The Planiverse: Computer Contact with a Two-Dimensional World,* just published by Pan Books, Dewdney describes how he and his students developed a close relationship with Yendred, one of the Ardeans, who explained to them many of the fascinating facets of life on a two-dimensional planet. Things that we take for granted would cause great problems for the Ardeans.

For instance, the Ardeans have an external skeleton, much like the insects here on Earth, since internal bones would obstruct the internal flow of "blood," etc. They have two arms on each side of their body, since it is impossible for them to bring a hand from one side of the body to the other.

Their alimentary canal does not pass all the way through their body, since that would cut them into two pieces: consequently their digestive system works differently from ours; as indeed do most most of their bodily functions. Their technology is also quite novel, though surprisingly advanced. In fact, the reader of Dewdney's book will be amazed at just how advanced a civilization can be when there are only two dimensions available.

Of course, following Dewdney's account of Yendred and his fellow Ardeans, skeptical readers will no doubt regard the whole thing as a work of science fiction. But then it was once regarded as nonsense to imagine that our world is round. And surely the onus is on the sceptic to read Dewdney's account and point to the flaw. They will have a hard time of it. In a technical appendix to the book, Dewdney collects articles written by various eminent scientists explaining the workings of the Planiverse. It is even explained how it is that the Planiverse could be "a part of" our universe without our ever being aware of its presence. I hope you enjoy your trip to Arde as much as I did.

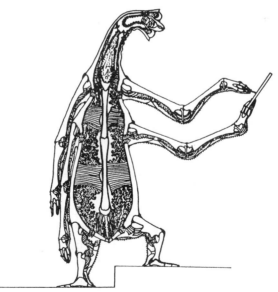

FIGURE 12.   A typical Ardean. This is not a cross-section of an inhabitant of Arde. The planet Arde is two-dimensional, so to three-dimensional creatures such as ourselves, looking down on Arde, its inhabitants really look like this.

23

How maths adds up to the best computer game of all

June 7, 1984

Why do mathematicians do mathematics? The answer will vary from individual to individual, but common to all will be the fact that mathematics is fun. To most people this marks out the mathematician as a rather odd character: any recollection of school mathematics is sufficient to send a shiver down the back. The truth is, it sends a shiver down mine as well. I only became a mathematician by accident. Initially I wanted to be a chemist, but the horrible pongs of some organic compounds soon sent me hurrying towards physics, an altogether cleaner subject, but one which required more mathematics than I had. In learning the mathematics I needed for physics, I discovered two things.

First, what I had previously thought was mathematics was really nothing more than glorified arithmetic, which I hate now as much as I ever did. Second, real mathematics, the kind mathematicians do, is fun.

I was lucky. The vast majority of people never get to that stage. Either they give up the subject before they really see what is involved, or else they adopt such a negative attitude that by the time they do meet some genuine mathematics they have lost all desire to try to understand it.

At my own university, as well as some others, we try to overcome the negative attitudes of the second of these two groups by hanging some mathematics onto the peg of necessity.

Students of any scientific (and several non-scientific) subjects at university level need mathematics to proceed. The ones with nothing more than an 0-level (and, not infrequently, not even that) form a captive audience for a special course which is designed to try and create some interest in the subject at the same time as providing the skills they need for their main subjects. As you might imagine, our success rate (in the sense of creating a genuine interest and love of mathematics) is not high, but every year we score some notable (and very rewarding to us) successes. Like all converts, we like nothing better than helping someone else to see the light.

Our task is made considerably easier by the widespread availability of cheap, efficient electronic calculators and computers, which free the student (and the teacher) from the unnecessary chore of doing sums. From this standpoint, the computer has been a great asset to the would-be mathematical evangelist. But that is not all. It does not require a great deal of research to discover that Britain has been gripped by micro-fever, especially amongst the young, and to the experienced evangelist, this huge army of micro owners is fertile preaching ground.

Once the novelty of playing packaged games has worn off, the enthusiastic micro owner starts to look around for something a bit more challenging to do with their expensive new toy. It was this state of affairs that led the *Guardian* to include this mathematics column on their computer page when it was launched last autumn. "Provide micro owners with some ideas," the editor told me. The type of letters I have received from readers since I started suggests that the editor was right in thinking there was a need for this kind of thing.

People who would have classified themselves as both useless at and uninterested in mathematics a few months ago are now getting embroiled in some fairly sophisticated genuine mathematics.

For instance, following a piece I wrote on Euler's prime generating polynomial, a reader in Bristol used his PC to work out an impressive theory of such polynomial formulas. No matter that as a trained mathematician I was soon able to develop his theory using elementary algebraic ideas. He did not have this knowledge, so to him what he did was nothing less than pure research, and good research at that. Perhaps he has even been motivated to try to learn the algebra involved.

24

How Archimedes number came up

June 21, 1984

How big is a million? I doubt if even those of us who have a million pounds can really appreciate the size of a million. For instance, suppose I agreed to give you one hundred pounds a week, every week of the year from now on, and that you decided to save each of my gifts in a suitcase until you had a million pounds there. How long would you have to wait? The answer is over 192 years. So perhaps I will not make the offer after all. In order to make you a millionaire in five years I would have to give you £3,846 a week, all year round.

When we hear about modern computers performing 400 million arithmetical operations in a single second, it puts things into some sort of perspective to remember that there are "only" 31,536,000 seconds in a (standard) year, and that 63 billion seconds ago (give or take the odd million) Christ was alive.

If a number with a paltry 11 digits can take us back to biblical times in seconds, what about numbers with hundreds or even thousands of digits? In modern cryptography, 200-digit numbers are used quite routinely now, and a computer has found the largest known prime number in the world, a number with just short of 40,000 digits. You may think that it is only quite recently that numbers of this

magnitude have cropped up, when computers have been around to help us. Not a bit of it!

In the third century B.C., the famous Greek mathematician Archimedes issued a challenge to the Alexandrian mathematicians, headed by Eratosthenes. Written in the form of an epigram, Archimedes's challenge begins thus: "Compute, o friend, the number of oxen of the Sun, giving thy mind thereto, if thou has a share of wisdom."

He then goes on to describe, in wonderfully poetic language, a certain herd of cattle, consisting of four types, with bulls and cows of each type. The number of cattle in each of the eight categories is not given, but these numbers are related by nine simple conditions which Archimedes spells out. For example, one of these conditions is that the number of white bulls is equal to the number of yellow bulls plus five-sixths of the number of black bulls. The problem is to determine the number of cattle of each category, and thence the size of the herd. (Actually, what is required is the smallest possible number, since the nine conditions do not imply a unique answer.)*

In his epigram, Archimedes goes on to say that anyone who solves the problem would be "not unknowing nor unskilled in numbers, but still not yet to be numbered among the wise."

---

* The complete problem is stated as an addendum to this chapter.

Nothing could be more apt, since 2,000 years elapsed before a computer finally found the solution. Clearly Archimedes had a mischievous streak in addition to his principles, and was trying to pull a fast one on his Alexandrian rivals.

In 1880, a German mathematician called Amthor showed that the total number of cattle in Archimedes herd had to be a number with 206,545 digits, beginning with 7766. Not surprisingly, Amthor gave up at that point. Over the next 85 years, a further 40 digits were worked out. But it was not until 1965 that mathematicians at the University of Waterloo in Canada finally found the complete solution. It took over $7\frac{1}{2}$ hours of computation on an IBM 7040 computer. After which no one thought to keep the printout of the answer! The world had to wait until the problem was solved a second time using a Cray-1 computer in 1981 for a published printout. It took the Cray-1 just 10 minutes to crack it. But after 2,000 years I think Archimedes has to have the last laugh.

Now you have been warned, here is a curious little problem posed, I think, by the famous English mathematician D.E. Littlewood. What is the smallest number such that when you take the first digit and move it to the end, the result is a number which is exactly 3/2 times the original number? If I were feeling really vindictive (like Archimedes?), I could just leave it like that, but for those of you who want to try to crack this on a micro, let me tell you that the answer is a number with 16 digits.

For the more mathematically minded amongst you, there is a cute little trick for finding the required number by mathematical reasoning. This trick, if not the original problem, is certainly due to Littlewood. Suppose the number is, in the usual decimal representation, $abcd\ldots xyz$. (So a, b, c, d, etc., are the digits of the number; we do not know how many there are at this stage.) Look at the fractional number which has the infinite repeating decimal expansion

$$a.bcd\ldots xyza\,bcd\ldots xyza\,bcd\ldots xyz\ldots.$$

(The first dot here is the decimal point. The others indicate missing digits.) From now on you are on your own.

I can think of no more fitting way to round off this week's column than with a further quotation from Archimedes. He ends his work, *The Sand Reckoner*, in which he presents a "rigorous" mathematical calculation that the number of grains of sand required to fill the "known universe" is (approximately!) 10 to the 63rd power (and which is incidentally the only mathematical work I know in which poppy seeds play a major part) with the words:

"I conceive that these things, king Gelon, will appear incredible to the great majority of people who have not studied mathematics, but that to those who are conversant therewith and have give thought to the question of the distances and sizes of the earth, the sun and the moon and the whole universe, the proof will carry conviction. And it was for this reason that I thought the subject would be not inappropriate for your consideration."

Reference

T. L. Heath, *The works of Archimedes,* Dover Publications.

ADDENDUM
Archimedes' Cattle Problem

"The cattle problem" has to do with a herd of cattle consisting of both cows and bulls, each of which may be white, black, yellow, or dappled. The numbers of each category of cattle are connected by various, simple conditions. To give these, let W denote the number of white bulls, w the number of white cows, B the number of black bulls, b the number of black cows, with Y, y and D, d playing analogous roles for the other colors. Using Archimedes' method of writing fractions (that is, utilizing only simple reciprocals), the first seven conditions which these various numbers have to satisfy are:

1. $W = (1/2 + 1/3)B + y$
2. $B = (1/4 + 1/5)D + Y$
3. $D = (1/6 + 1/7)W + Y$
4. $w = (1/3 + 1/4)(B + b)$
5. $b = (1/4 + 1/5)(D + d)$
6. $d = (1/5 + 1/6)(Y + y)$
7. $y = (1/6 + 1/7)(W + w)$

The two remaining conditions are:

8. $W + B$ is a perfect square (that is, equals the square of some number)
9. $Y + D$ is a triangular number (that is, is equal to a number of balls that may be arranged to form a triangle, which is the same as saying that the number must be of the form $n(n + 1)/2$ for some n).

The problem is to determine the size of the eight unknowns, and thus the size of the herd. More precisely, the aim is to find the *least* solution, since the conditions admit more than one solution.

If conditions (8) and (9) are dropped, the problem is relatively easy. The smallest herd consists of 50,389,082 cattle. The additional two conditions make the problem considerably harder. It has been claimed that the first complete solution was worked out by the Hillsboro (Illinois) Mathematics Club between 1889 and 1893, though no copy of their solution has been found, and there is some evidence to suggest that what they in fact did was work out some of the digits of the 206,545-digit solution and provide an algorithm for the computation of the remainder. In 1965, H. C. Williams, R. A. German, and C. R. Zarnke at the University of Waterloo in Canada used an IBM 7040 computer to crack the problem once and for all. The final solution occupied 42 sheets of printout. In 1981, Harry Nelson repeated the calculation using a Cray-1. This machine took a mere 10 minutes to come up with the answer. Reduced to fit 12 pages of printout on a single journal page, the solution was published in the *Journal of Recreational Mathematics* 13 (1981), pp. 162–176.

25

Biblical fingers get stuck into pi

June 21, 1984

Noting that there are two references to the mathematical constant π ("pi" = 3.14159...) in the Bible (I Kings 7:23 and II Chronicles 4:2), both of which imply the value $\pi = 3$, some academics in Kansas have started The Institute for Pi Research, whose avowed intention is to propagate the usage of the value 3 for π. As the Institute's founder, Samuel Dicks, Professor of Medieval History at Emporia State University, says: "If a π of 3 is good enough for the Bible, it is good enough for modern man."

The Institute is campaigning for the value $\pi = 3$ to be given equal time with the more conventional value in state schools. Coupled with Dick's statement that: "I think we deserve to be taken as seriously as Creationists," this would appear to give some clue as to what really lies behind the foundation of the Institute. As economic historian Loren Pennington says, "If the Bible is right in biology, it's right in math."

But whatever their real aim, they seem to have friends in high places. The group wrote to President Reagan asking for his support, and though they did not receive a reply, they were greatly encouraged to hear him say in a speech shortly afterwards that: "The pi(e) isn't as big as we think."

26

The hidden holocaust

July 5, 1984

High level computer languages such as BASIC, Pascal, or Fortran enable computer users to write programs in a reasonably understandable fashion. Pascal, in particular, results in programs which can almost be read like a book. But any such language imposes quite severe restrictions on the computer user; it is only possible to do the kind of computation for which the language was designed, often file handling and arithmetic, with possibly graph plotting thrown in as well. (Computer graphics do not really form a part of, say BASIC, though for ease of use most micro manufacturers incorporate graphics commands in their implementation of the language.) Consequently, ambitious micro owners usually resort to writing all or part of their more advanced programs in Assembly Language.

Assembly Language is, to all intents and purposes, the numeric language in which the computer itself operates. An Assembly Language instruction such as

ADD(A,B,C)

would represent the single computer operation of adding the contents of memory locations A and B and putting the result in location C. It would not matter if A and B contained not numbers but letters; to the computer these are indistinguishable, the difference being recognized only by high level languages like BASIC. Because Assembly Language allows the programmer to control the operation of the computer at its most basic level (almost), it puts no restriction upon what can be done. For instance, an Assembly Language program may contain instructions which modify itself, or which replicate itself, or even destroy itself, all of which are explicitly prohibited by most high level languages.

A story that did the rounds some years ago claimed that some malicious (or do I mean mischievous) company programmer wrote an Assembly Language program which did nothing other than replicate itself, amoeba style, gradually filling all the storage space not only in his firm's mainframe computer, but all the memory in the company's entire computer network. I have no idea if this is true or not, but something of that nature is not at all inconceivable.* This story is repeated in an article by A.K. Dewdney in the May issue of *Scientific American.* (Regular readers of this

---

* In fact, this did subsequently occur. In 1988, the so-called "Internet worm" brought the North American academic computer network to a halt. In fact, the proliferation of 'computer viruses' became such a problem in the latter part of the 1980s that most present-day computer owners/operators employ special software routines to protect their systems from damage.

column have come across Dewdney before in connection with his novel *The Planiverse* about life on a two-dimensional planet. [See Chapter 22.])

In his article, Dewdney describes a computer game called Core War, which he developed with one of his students at the University of Western Ontario in Canada, where he is Professor of Computer Science. The game is played by two players, or rather by two computer programs. The two participants each write a computer program in a specially tailored form of Assembly Language. This language does not allow them to refer to specific locations in the computer memory, only to "relative addresses," which means it is permissible to refer to a memory location indirectly, say 100 locations beyond the current instruction, but not to, say, "location 4083." The two programs are loaded into the computer under the control of a master program, which then operates the two programs one instruction at a time in turn. The initial location of the two programs is chosen by the master program, so neither program knows where the other is.

What is the point of all this? Survival. The object is to write a program which can destroy the opponent (by making it inoperative) while evading destruction itself in the process. How this is done is entirely up to the programmer. He may write a program which moves through the memory very fast, overwriting all in its path, in the hope of killing his opponent in the process. Or he may include a routine for sending out a stream of little bomber rou-tines to hunt out and destroy. Defensive tactics could include setting up a ring of guard-posts around the program, or incorporating some self-repairing mechanism, like making a copy of the initial program and transferring control to it if the original is hit.

The only restriction on the tactics are those imposed by the limitations of the programmer's ability.

The aspect of this that I find particularly fascinating is that the actual combat between the two programs proceeds unseen and unheard, deep in the core of the computer. (Hence the name Core War. The name "core" for the central memory of a computer goes back to the early, pre-chip computing era, by the way, when the central memory was constructed from ferromagnetic cores attached to a grid of fine wires.)

There is, of course, no inherent reason why Core Wars could not be extended to allow for more than two players. Nor do they have to be located at the same place. Computer networks could be used to wage wars across nations or even between continents. The winner would be the one who developed the best strategy for both defense and attack. In other words, the best technologist would win. Rather like the view of global conflict held, it appears, by many of the world's leading politicians. But at least in this version the loser would lose nothing more than his pride, and the winner would still be around to enjoy the victory, which is hardly likely to occur in the real counterpart of such a conflict.

27

All in the mind

July 19, 1984

There is little doubt that the computer revolution has led to a much wider interest in mathematics. This is partly because many micro owners find themselves learning more and more mathematics in order to get the best from their machine. (For instance, you will not get far with a graphics package unless you learn about Cartesian geometry.) But much more significant is the fact that the computer, and more specifically the pocket calculator (which, remember, is simply a microcomputer with a few built-in subroutines and, in general, no programming features) frees the would-be mathematics student from the chore of arithmetical calculation.

Many people confuse mathematics with numerical calculation, though in reality this is akin to confusing an architect with a bricklayer. (In fact, relatively few mathematicians have any professional contact with "numbers" at all, at least in the sense of performing calculations.) Since there is no possibility of the pocket calculator being disinvented, mankind has been freed forever of the need to perform tedious calculations using pencil and paper. For simple tasks, a calculator will do the job more accurately and efficiently, while the computer is ideally suited to more complex numerical manipulations.

For most of us, this development is a tremendous improvement. We can concentrate on what is involved in solving any given problem without getting sidetracked into arithmetical details. Also the computer can be of great assistance for the research mathematician. If a certain mathematical result is suspected, it is in many cases possible to use the computer to gather evidence for that result, evidence which can later lead to a complete mathematical solution of the problem.

But is mankind going to pay a price for the reliance on machines to perform calculations? No, this column is not being ghost-written by a retired mathematics master and one-time colonel living in Tunbridge Wells. I am not trying to suggest that mental arithmetic is somehow good for the soul, or that the computer will lead to our eventual downfall. But for a very small number of highly gifted mathematicians, the intense familiarity they developed with numbers through calculation has led to some profound mathematical discoveries.

For instance, the great Indian mathematician Ramanujan, upon being told by his friend and colleague the Cambridge mathematician G.H. Hardy that he had just ridden in taxi-cab number 1729, observed at once that this was the smallest number which is expressible as a sum of two cubes in two different ways. Now in itself this is not particularly deep, though it is closely related to some extremely deep and still not fully understood mathematics, not without application in the real world.

But imagine the familiarity with numbers that must be required in order to spot something like this. Of course, with a microcomputer it is an easy task to check Ramanujan's claim, once you know what it is. But Hardy did not simply ask Ramanujan if the result was true. He gave him the number and asked if there was anything special about it. Imagine doing that on a computer: give it a number and ask if it has any unusual properties. (Incidentally, lest it be thought that Cambridge dons spend all of their time asking each other silly questions like this, I should add that Ramanujan was ill at the time and Hardy had visited him by taxi to try and take his mind off things and cheer him up a bit. Cambridge dons really only spend *part* of their time asking each other stupid questions.)

To take another example, would Fermat (1601–1655) have obtained all of his profound results about numbers had he not learned to regard them all as intimate friends as a result of long-hand calculations? Or Gauss, or Euler, or all the other great 18th century mathematicians? These giants made deep mathematical discoveries simply in order to be able to perform calculations with very large numbers, and practically all of the present day computer algorithms used to handle large numbers in cryptography, research on large prime numbers, calculating π to many decimal places, and so on, depend upon mathematics developed by these grand old men. Given the technology, any one of these mathematicians could have obtained practically all of our present day computer discoveries over a hundred years ago.

28

Find a four-letter word and the square root of computer
August 6, 1984

Take the word "computer." Assign a digit to each letter in this word, different digits to each letter. By replacing each letter by its digit you will obtain an 8-digit number. This number may or may not be a perfect square. If it is, take its square root. This will be a 4-digit number. Its digits may or may not be distinct digits which are among those assigned to the letters in "computer." If they are, it may or may not be the case that when you replace the four digits in the number by the appropriate letters, you end up with a word of the English language.

In fact there is only one way of assigning digits to letters that makes this possible. Curiously enough, the four-letter word that results is one which is familiar to all computer programmers.

What is this four-letter word? Your micro should be kept busy for quite some time sorting this one out.

If that is not enough for you, how about this one? If you take the two 2-digit numbers 21 and 87 and multiply them together the answer is 1827, a 4-digit number which consists of the same four digits that make up the original two numbers. How many other examples of this phenomenon can you find? Again this is a problem which is well suited to computer attack.

While your micro is busily whirring away with the above problems, here is one to keep your original, biological micro at work. It was devised by my Lancaster colleague Tony Llewellyn, senior lecturer in computer studies and former international athlete.

Big Brother was reviewing the progress of three of his agents, who prefer to be referred to as *A*, *B*, *C*, to protect their anonymity. These three had been sent out to sell copies of *Brainspeak Weekly*. Big Brother pointed out that although he judged all three to be equally competent in their logical analytic capabilities, their performances as newspaper salesmen were different.

A had commendably sold more copies than *B*, who had sold more copies than *C*. However, he noted, the product of the number of copies sold was a significant number, namely 1984. All three scribbled furiously for a few moments in an attempt to see how many copies each had sold, after which the following conversation took place. Big Brother asked *A* how many copies *C* had sold, to which *A* replied that he did not know. Big Brother then asked *B* how many copies *A* had sold, and *B* said he did not know. Asking *C* how many copies *B* had sold, Big Brother again received a "don't know" reply, as he did when he asked *A* once more after the number of copies sold by *C*. If Big Brother now asks *B* how many copies *A* sold, what should *B* reply (assuming *A* wishes to impress Big Brother)?

29

Question time

August 16, 1984

This week's column is devoted to a selection of holiday season teasers which can be solved without the aid of a micro. Like all good tests, there are some "what do you know" bits and some "can you solve it" bits. This being your super soaraway *Guardian,* there are no prizes to be won, but if you get them all right you can at least lie back on the beach with a smug, self-satisfied feeling. The answers are given on pages 70–71.

1. On June 21, 1948, the first computer program written for a stored program electronic computer was successfully run. Who wrote it and what did it do?

2. She was titled, a compulsive gambler, an alcoholic, an opium addict, and (it is alleged) somewhat promiscuous. Her father was a famous poet. She was honored some years ago by the United States Defense Department. Who is she, and what is she doing in *Micro-Guardian*?

3. Why is the computer language BASIC so named?

4. Once upon a time, around A.D. 760 in fact, there was an Arabian mathematician called al-Khowarizmi. Why could it be said that his name is practically a household word in computing circles?

5. Most modern computers have a particular location in their central processing unit known as the B-register. How did this come to be so named?

6. Regular readers of this column will know that supercomputers are often used to find record prime numbers, the current holder being a 39,751-digit prime discovered in September 1983 using a Cray XMP computer. When were electronic computers first used to search for large primes?

7. Hermann Hollerith worked for the United States Census Bureau in the second half of the 19th century. He had the clever idea of using punched card technology (then used for weaving) to help tabulate the census data. In 1896, Hollerith formed the Tabulating Machine Company to capitalize on his idea. By 1911, after one or two mergers, the company was known as the Computing-Tabulating-Recording Company. What became of that company?

8. Most computer keyboards use the so-called QWERTY keyboard, so called because the top row of keys consists of the letters QWERTYUIOP. Can you make a 10 letter word using these letters (repetitions allowed)?

9. Employees of the Bell Telephone company in the USA are reputed to have manufactured lapel badges with the legend "Ma Bell Keeps a Baudy House." What is a Baud?

10. What does GIGO mean?

11. If you take the digits 1 to 9 in order, there are exactly 11 ways in which you can insert plus and minus signs to give a sum with answer 100. One of these is

$$123 - 45 - 67 + 89 = 100$$

What are the other 10?

12. Find digits A, B, C such that $ABC = A! + B! + C!$ (The exclamation sign here is the mathematician's sign for the 'factorial function.' For any number N, $N!$ denotes the result of multiplying together all of the numbers 1 to N inclusive.)

13. Find the four numbers which are equal to the sum of the cubes of their digits. (Exclude the trivial case of 1.)

14. Any mathematician will tell you that it is impossible to square the circle. At least, in the old Greek tradition that is the case. But how about "squaring the circle" in the following sense: replace each of the asterisks in the figure below with a letter so that each row and column of the resulting square is an English word.

```
  C  I  R  C  L  E
I     *  *  *  *  *
R     *  *  *  *  *
C     *  *  *  *  *
L     *  *  *  *  *
E     *  *  *  *  *
```

15. The following describes a unique number. Which number? It is not divisible by the square of any prime; and for any prime p, p divides the number if, and only if, $p - 1$ divides the number.

That's it. Have a nice day.

Final Score

1. The program was written by T. Kilburn, and its aim was to find the highest factor of a given whole number. It was written for, and run on, an experimental computer built at Manchester University. (Earlier "computers" such as the Colossus, used in wartime code-breaking, were not "stored program" machines.)

2. This amazing lady was the Lady Augusta Ada Lovelace (1816–1852), daughter of Lord Byron. A collaborator of Charles Babbage, the inventor of two famous mechanical "computers," Lady Lovelace developed the basic principles of computer programming, including the concepts of loops and subroutines. The US Defense Department named its programming language ADA after her.

3. It stands for Beginners All-Purpose Symbolic Instruction Code.

4. The word "algorithm" is a derivation of the name al-Khowarizmi. Al-Khowarizmi wrote a book outlining the rules for performing arithmetic with numbers written in the decimal form used today.

5. The memory of the first ever electronic computer, the Manchester machine, consisted of cathode ray tubes. Initially there was an arithmetic tube, called the A-tube, and a control tube, the C-tube. What could be more natural than to call the next one, introduced later, the B-tube.

6. The Manchester computer again. In June 1949, following a suggestion of Max Newmann, a search for Mersenne primes was carried out up to the 353rd Mersenne number. The search was unsuccessful, since the next Mersenne number which is prime is the 521st.

7. In 1924, by then under the control of one Thomas J. Watson Sr., the company was renamed International Business Machines: IBM for short.

8. Using the top row of a standard typewriter keyboard you can spell out the word

TYPEWRITER.

9. A baud is a unit for measuring the rate of transmission of data: 1 baud is a transmission rate of one bit per second.

10. GIGO stands for "Garbage in, garbage out," a phrase that should be etched into the heart of all computer users, especially those employed in Government service. Computer output is only as good as the input data allows.

11-15. Sorry. You should know me by now. I'll leave you to sweat over these mathematical ones for a while, and come back to them at a future date. [See Chapter 31.]

30

First find your algorithm

August 30, 1984

The rules for performing basic arithmetic using numbers expressed in the Hindu decimal form that we use today (with columns for units, tens, hundreds, etc., and decimal points to denote fractional quantities were outlined in a book written some time around A.D. 760, by an Arabian mathematician called al-Khowarizmi. From his name comes the modern word *algorithm,* which is much heard in computer circles.

What exactly is an algorithm? Briefly, it is a set of instructions that must be followed in order to solve a specific problem, usually in mathematics, though this is not necessarily the case. The simplest example is one described by al-Khowarizmi himself, and familiar to anyone over seven years of age. When we are taught the rules for adding two numbers, we are learning an algorithm. In this case, the algorithm tells us to write one number beneath the other, with any decimal points lined up correctly and to start adding column by column from the right-hand end, carrying into the next column to the left whenever the answer is greater than 9.

The important point about this procedure is that it works for any pair of numbers. A rule that only worked in a specific case would not be called an algorithm, but a general and universally applicable rule would be so called. For instance, instructions as to how to change a tire on a car constitutes an algorithm.

A computer program is simply an algorithm expressed in some specific computer language such as BASIC, Fortran, or Pascal.

Once an algorithm has been worked out for solving a given problem, writing it in the form of a program is relatively easy. Being essentially a linguistic task, what is required is a familiarity with the programming language, together with a certain amount of experience, but, except on rare occasions, no great knowledge of, or flair for, mathematics. Finding the algorithm in the first place is, on the other hand, much more difficult.

Indeed, it is known that there are mathematical problems for which no algorithm exists nor ever can exist. Such problems could never be solved using a computer. The only chance of solution is by some stroke of genius on the part of a mathematician. Even when a problem can, in principle, be solved by means of an algorithm, that is often not the end of the matter. The question then arises as to whether it is feasible to use that algorithm as the basis of a computer program. What is at issue is how long it would take the computer to solve the problem if it were programmed to use the algorithm. Many simple arithmetical problems can be solved using "obvious" algorithms which turn out to require vast amounts of expensive computer time to run. In such cases the mathematician seeks a more subtle, and hopefully more efficient algorithm.

The multiplication of two large numbers serves as a good illustration. Suppose we want to multiply two numbers, each involving N digits. If we write a program to do this using the standard algorithm we all learn at school, this involves some $N \times N$ individual multiplications of digits, since we must multiply each digit in one number by each digit in the other in turn. Since the multiplication of two digits is performed in a time which depends only upon the computer, mathematicians would say that this algorithm "runs in time" $N \times N$.

By and large, even for large values of N, say $N = 100$, this is a perfectly feasible algorithm for a modern computer, which is capable of performing more than one million basic arithmetic operations per second. But since multiplication of numbers is such a common operation, it would be advantageous to find an even more efficient method. Surprising though it may seem to us, having been brought up on the standard school algorithm, there are algorithms which are much more efficient indeed.

The fastest multiplication algorithm that I know of was developed by V. Strassen and A. Schonhage in 1970. It multiplies two N-digit numbers in just over N steps. The precise running time is N times the number of digits in N times the number of digits in that last number. For instance, suppose we want to multiply two numbers of 1,000 digits (an exceedingly rare occurrence). The number 1,000 has 4 digits. The number 4 has 1 digit. So the multiplication runs in time $1,000 \times 4 \times 1$, that is in time 4,000. The "classical" algorithm runs in time 1,000,000 which means that the new algorithm is some 250 times faster than the old one in this case.

The Strassen–Schonhage algorithm achieves this incredible speed by utilizing some fairly sophisticated mathematics, and this column is not an appropriate place to describe it. Anyone sufficiently interested to want to see the gory details should look at pages 270–274 of Donald Knuth's book *The Art of Computer Programming,* Volume 2 (Addison-Wesley).

The problem which is probably receiving the most attention just now, partly because of its connection with coding theory, is that of finding an efficient algorithm for finding factors of a given number. Methods for doing this other than by trial and error do exist, though they are not taught in schools. But there are none known that are at all efficient.

Finally, and unrelated to the above, a reader has asked me about the following problem. You have to find prime numbers with the property that all the numbers you get by chopping off digits from the end are prime. For instance, 19139 is prime, and so too are the numbers 19, 191, and 1913. His question is: "Have I come across this before?" to which the answer is "no." So I am passing it on to you. Does anyone know anything about it? Are such prime numbers common, and if so just how common? Are there any theories about it?

Obviously, the problem is ideal for a computer attack, and rather than look at it myself I shall leave it to you to have a go and (hopefully) let me know how you get on.

Postscript.   See Chapter 36 for an update on the last problem.

31

Circle games

September 13, 1984

Most of this week's column is devoted to a roundup of responses to the quiz I set on August 16 [Chapter 29], but first let me answer a reader's question concerning the origin of a term used by computer professionals.

Professional programmers often refer to textual output from a computer (as opposed to numerical output) as "Hollerith strings." Indeed, in the computer language Fortran, the letter H is used to indicate text input and output. For example, the Fortran instruction

```
WRITE8H Guardian
```
would cause the output of the string "Guardian." (The 8 refers to the number of symbols in the string.)

Hermann Hollerith worked for the US Census Bureau in the second half of the 19th century, and had the idea of utilizing for the handling of census data the punched card technology invented by the French weaver Joseph Jacquard for weaving programmed patterns into cloth. The idea proved to be so successful that the punched card soon became known as the Hollerith card. In 1896, Hollerith formed the Tabulating Machine Company to capitalize on his invention. By 1911, as a result of mergers, the company became known as the Computing-Tabulating-Recording Company.

Enter a certain Thomas J. Watson, Sr. In 1914, after 20 years with National Cash Register as a salesman, Watson was fired, and be-

came president of the Computing-Tabulating-Recording Company. In 1924 he decided to rename the company, choosing the impressive title "International Business Machines"; IBM for short. The rest, as they say, is history.

On August 16, by way of a change from the usual fare, I set a summer quiz. One question in particular produced a tremendous response from readers: squaring the circle. As posed by the ancient Greeks, this problem asks for a square to be constructed with area equal to that of a given circle. The only tools to be used are the classic Greek ones of ruler (more precisely, a one-sided, unmarked straightedge) and compasses (of a type which lose their separation when taken off the paper). In 1882, Lindemann proved that the mathematical constant π cannot be the solution to any polynomial equation with whole number coefficients, a result which easily implies that the problem of squaring the circle cannot be solved. The problem I asked was whether you could square the circle in the square

```
C I R C L E
I * * * * *
R * * * * *
C * * * * *
L * * * * *
E * * * * *
```

so that every row and every column spells out a word.

The most common solution I received was:

```
C   I   R   C   L   E
I   N   U   R   E   S
R   U   D   E   S   T
C   R   E   A   S   E
L   E   S   S   E   E
E   S   T   E   E   M
```

Also common were:

```
C   I   R   C   L   E
I   B   E   R   I   S
R   E   C   E   S   S
C   R   E   A   T   E
L   I   S   T   E   N
E   S   S   E   N   E
```

and

```
C   I   R   C   L   E
I   C   A   R   U   S
R   A   R   E   S   T
C   R   E   A   T   E
L   U   S   T   R   E
E   S   T   E   E   M
```

Dominic Francocci from Leamington Spa managed to come up with nine different solutions, one of which used the word *cledge,* which I pass on to keen Scrabble players. In fact, with words like *resect, lintie, isohel, chasse,* and *litten* cropping up, I suspect that quite a lot of you are Scrabble buffs. The greatest number of solutions I received was 55, sent to me by Emrys William in Milton Keynes. How did he find so many? You've guessed it. He used a computer. Supplied with the 2,516 words of six letters beginning with any of c, i, r, l, e listed in Chambers *Words,* the computer took no time at all to produce the goods.

I am tempted to ask you to try to cube the cube in the obvious way, but I gather that the circle squaring problem caused several marriages to come close to breaking point, so perhaps I won't.

Other problems. The three digits A, B, C which give $ABC = A! + B! + C!$ are 1, 4, 5.

The only four numbers equal to the sum of the cubes of their digits (excluding the trivial case of 1) are 153, 370, 371, and 407.

The last problem caused some headaches, I gather. You have to find the unique number which is not divisible by the square of any prime and is such that for any prime p, p divides the number if, and only if, $p - 1$ divides the number. The trick is to find just which primes divide the number. Since 1 divides the number, the prime 2 will. Hence so also will 3. 5 does not (since 4 doesn't), but 7 does (since 6 does). The only other one is 43. So the number is 1806 (i.e., $2 \times 3 \times 7 \times 43$).

Finally, the answer to the Big Brother problem posed on August 2 is that A sold 124 copies of *Brainspeak Weekly.*

32

How the Babylonians almost saw the point

September 27, 1984

In an earlier article [Chapter 5] I mentioned the introduction of the binary system of numbers for use in electronic computers, explaining how the construction of a modern computer practically dictates that all numbers be represented internally in binary form (i.e., using the two binary digits, or "bits," 0 and 1 only), even though all communication between the computer and the outside world may involve the more standard (for us) decimal notation, using the ten digits 0 through 9.

Thus, for example, when we feed the number 13 into a computer, this is stored internally as 1101, which is the binary representation of the same quantity. (In binary, instead of columns for units, tens, hundreds, etc., there are columns for units, twos, fours, eights, sixteens, etc. Arithmetic in such a system is performed in the same way as for decimal notation, except that there is a carry whenever there is a multiple of 2 or 4 or 8, etc., rather than for a multiple of 10, 100, 1,000, etc.)

Just as the binary system is the most convenient for a computer to use, so the decimal system is the most convenient for a human being counting on his fingers. But in spite of this "advantage" of the decimal system, mankind has not always made use of it.

In some primitive cultures we can still see use being made of the earliest forms of number representation. Usually these are based on groups of fingers, piles of stones, scratches on wood, and so on, with special conventions for replacing a larger group or pile by a single special object or an object in a different place. These primitive systems lead naturally to the earliest ways of writing numbers, such as the systems of the Babylonians, Egyptians, Greeks, Chinese, and Romans. But none of these systems are suited to performing arithmetic, except in the most simple of cases.

The Babylonians actually had two distinct systems for representing numbers, 2,000 years B.C. Their business system involved writing numbers in a form of decimal notation, using groupings of tens, hundreds, etc. Large numbers were seldom involved in calculations, so their system, which was adapted from earlier ones from Mesopotamia, served them adequately. But for mathematical and scientific work, the Babylonians used a sexagesimal positional notation. That means that the "base" used was 60, and the *position* of a digit indicated whether the digit represented a unit, a multiple of 60, a multiple of 3600, or whatever. (So they had a units column, a 60s column, a 3600s column, etc.)

This type of system is still in use today, of course. We measure angles using such a system; 60 seconds equal one minute, 60 minutes equal one degree, and, a slight departure from this nice pattern, $360 = 6 \times 60$ degrees gives a full circle. Likewise time is measured using units of 60.

Today, in using the decimal system, we use a decimal point to indicate fractional quantities, so that 356.78 equals 100 times 3.5678, etc. The explicit placing of a decimal point in a number like this gives rise to what is known as *fixed-point representation*. In performing arithmetic on such numbers, we keep track of the position of the point throughout the calculation.

There is an alternative method, and that is to suppress mention of the decimal point within the number itself, and keep a separate record of where it should be. For instance, we could write the number 356.78 as 35678;2, where the final 2 indicates that there are two digits to the right of the point. This gives rise to what is known as *floating-point arithmetic*. The sexagesimal system used by the Babylonians was a floating-point system, except that all mention of the "position of the sexagesimal point" was omitted, the context of the problem indicating this quantity.

Modern computers also use floating-point representations of numbers involving fractions. As an illustration, consider the machine that I am most familiar with, the VAX-11 computer. It is capable of handling numbers whose binary representation has up to 32 bits. (By comparison, a pocket calculator or a typical home computer can handle numbers ex-

actly half this size whereas the giant super-computers such as the Cray-1 or the Cyber-205 allow for numbers with 64 bits.)

When handling numbers in floating-point binary form, the 32-bit VAX word is divided up into three pieces. One bit is used to tell whether the number is positive or negative; 8 bits are used to keep track of the binary point; and the remaining 23 bits carry the number itself (minus any binary point.)

The main drawback with floating-point arithmetic on a computer is that, in order to allow for a vast range of values, some accuracy is sacrificed. For the VAX-11, if a number requires more than 23 bits to represent it, it is only possible to store that number approximately in the computer. This has the effect that some of the properties of arithmetic that we take for granted are no longer valid when carried out on the computer. For example, it is no longer true that $(a \times b) \times c = a \times (b \times c)$ for any numbers a, b, c.

This means that the computer programmer has to be careful in the way the programs are written. High-school arithmetic cannot be totally relied on. Indeed, when floating-point arithmetic is concerned, it is true to say that a modern computer is only as good as its programmer.

Postscript.   See also Chapter 34.

33

The best way to get from A to B is by way of C and D

October 6, 1984

The recent dramatic rises in petrol costs, bringing the £2 gallon ever closer, will have encouraged many motorists to start (or restart) some sort of car-pooling scheme. The reason for the parenthesized caveat is that, as many a seasoned car-pooler knows, finding a pooling schedule that everyone agrees is fair is by no means easy, and car-pools have a habit of breaking up under the strain of trying to fit everyone's requirements. What would be nice would be some mathematically proven method for regulating the pool. Well, there is one.

In a paper entitled "A Fair Car-Pooling Scheduling Algorithm," published in the *IBM Journal of Research and Development* last year, IBM mathematicians Ronald Fagin and John Williams looked into the whole question of pooling schedules, examining various schemes as to both their fairness to all participants and their "robustness" in being able to cope with everyone's needs and whims.

The simplest method of organizing a pool is, of course, to change the driver on a daily or weekly basis. The problems with this scheme are manifest. To run smoothly it requires all participants to travel every day, and the driver for that day to be able to fulfil that role, regardless of the fact that said driver's spouse may well have other plans.

To inject some degree of flexibility into the scheme, most groups resort to "swapping"

driving days. But after a while this becomes so involved that no-one is quite sure where they stand (or sit), and begin to feel hard done by.

Another popular and attractive looking method is to introduce some kind of token into the pool. Each time someone travels in a car as a passenger, they give the driver one token. The driver is chosen to be the one able and willing to drive that day having the least number of tokens. This is also not a fair system, of course, since there is a considerable advantage to be gained from being the driver on a day when the entire pool is in the car, and a considerable disincentive to drive with only one passenger.

The problems with the two above schemes can be overcome by making separate record cards for every possible combination of car occupants, and recording on the appropriate card the driver that day. Assuming that there are four persons in the pool, Arthur, Bill, Charles and David (A, B, C, D for short), and that one day A, B, and D wish to go in the car, they look at the card for A, B, D and see who has the least number of "ticks." That person drives and enters one tick against his name. This scheme does seem to be fair, and indeed can be shown mathematically to be fair (see below), but it does have a rather obvious disadvantage. For four participants it requires 15 separate cards to cover all possible groupings of people. (For five participants, 31 cards are required.) Most poolers would regard this as just a bit too much.

In the last paragraph I talked about a pooling being provably fair. But what exactly does this mean? In their paper, Fagin and Williams give a precise mathematical definition of "fairness," which corresponds to the "obvious" notions of fairness to all participants, and are able to prove that the above scheme does satisfy these fairness criteria. So, too, does the method I shall describe last of all, the method the authors recommend as a result of their study.

Suppose there are N members in the pool. Let U be the least common multiple of $1, 2, 3, \ldots, N$, i.e., the smallest number into which each of $1, 2, 3, \ldots, N$ can be divided. This quantity U is the nominal "cost of a single journey by car" in the pool. The idea is that, for each trip, U is divided evenly between each person in the car (including the driver). Each passenger "pays" the driver his portion of the "cost." Of course, this "payment" would probably take the form of an entry in a book. In which case, the simplest method would be to allocate each pool member a column, and make the daily entries row by row. If there are K people in the car on a certain day, the driver adds $U \times (K - 1)/K$ to the total in his column up to that day, while each passenger subtracts U/K from their column. The driver each day is the person with the lowest number of "points" among the potential drivers for that day.

For instance, suppose that our pool consists of A, B, C, D from before. They begin the pool by entering zero in each column. Suppose that on day 1, B drives and C and D are passengers. In this case, with 4 participants, U is 12, so B enters 8 in his column while C and D each enter -4. A still scores 0. On day 2, D elects to drive (if possible, it ought to be either C or D since they both have the lowest scores to date), and all four members travel in the car. D gains 9 points, making his total 5, while A, B, C each score minus 3 points, making their scores -3, 5, and -7, respectively.

On day 3, C should drive, if possible. Suppose that he does, and that A is the only passenger. Then A's score goes down to -9 while C's goes up to -1 (a transfer of 6 from A to C), and the scores of B and D remain the same as before. And so on. So for the first three days of the scheme, the table would look like this:

	A	*B*	*C*	*D*
Day 0	0	0	0	0
Day 1	0	8	−4	−4
Day 2	−3	5	−7	5
Day 3	−9	5	−1	5

On day 4, A should drive, if possible; failing that C; and failing that B and D may toss a coin to decide who should drive the other.

Simple? Certainly. Moreover, this method has been shown mathematically to be fair to all the participants.

34

On making arithmetic pointless

October 11, 1984

The largest number which can be stored in a typical home computer is 65,535, the number which consists of sixteen ones in binary notation. For a 32-bit mini-computer the largest number which can be stored is 2,147,483,647 (one bit being reserved for the sign of the number in such machines). In a 64-bit mainframe computer, the range goes up to more than 18 million million million. In order to handle numbers larger than this you have to resort to programming devices which split the number into smaller parts stored in several locations in the computer. Software algorithms for performing arithmetic involving such "multiword" numbers have been developed; they are known as multi-precision arithmetic algorithms, and some of them operate very efficiently, but, since they inevitably require a number (often a quite large number) of basic machine operations in order to achieve a single multiword operation, they are nothing like as fast as the basic operations supplied by the hardware.

So far, everything I have said refers to so-called *integer* arithmetic, where whole numbers are involved and a perfectly accurate answer is required. But most practical uses of computers do not involve either of these constraints. Problems arising in real life usually involve fractional quantities, and only require an answer to a specified degree of accuracy. In order to handle numbers in this kind of situation, computers are designed to be able to store and handle numbers in what is called "floating-point" form.*

Briefly, this means that the computer word storing the number is divided (or rather acts as if it were divided) into two parts. One part stores the number, or at least as much of the number as is possible, without any decimal points (or binary points) or end zeros, while the other part stores the position of the decimal (or binary) point. For example, the number one million could be stored as (1; 6), the number 0.03 as (3; -2). (I am simplifying things, but this is the general idea.)

Arithmetic with numbers in floating-point form is usually referred to as *real* arithmetic. Using floating-point form, the size of number which can be handled in the computer increases dramatically. For instance, in a 32-bit computer, the range of numbers which can be stored goes up to around 10^{38} (that is a 1 followed by 38 zeroes).

If you want to perform arithmetic involving numbers even larger than those permitted by

---

* See also Chapter 32.

the computer's floating-point hardware, you again have to resort to programming techniques (multi-precision floating-point arithmetic), with the corresponding loss of speed. If you do not, and simply write your program to use standard computer operations, then if the numbers involved in the resulting calculation become too large, the result is what is known as "floating point overflow." Put simply, the numbers just overflow the computer's capacity and, depending on the computer and the programming language being used, either the program stops or, much worse, continues on with what is now a totally meaningless calculation.

The computer programmer has to be aware of the ever present possibility of floating-point overflow in the course of a calculation, particularly a long calculation involving large "loops." Indeed, studies have indicated that the possibility of overflow can be quite high.

It would be nice to have a system of representing numbers in a computer which was not subject to the problem of overflow. Well, it seems that one is on the way. Developed by Charles Clenshaw of Lancaster University and Frank Olver of the University of Maryland, and described in their paper "Beyond Floating Point" in the *Journal of the Association of Computing Machinery,* Vol. 1 (1984), pp. 319–328, the system represents numbers in what they call *level-index form.*

The idea behind this is not new. Mathematicians have known for centuries that you can represent enormously large numbers by iterating the exponential function. For instance, a few moments spent in front of a home micro should convince you that the sequence of numbers $\exp(0.5)$, $\exp(\exp(0.5))$, $\exp(\exp(\exp(0.5)))$, etc. rapidly becomes very large.

What Clenshaw and Olver have done is develop methods for performing arithmetic on such iterated exponentials in an efficient manner, a task which requires considerable mathematical ingenuity.

In level-index form, any number is represented as the result of iteratively applying $\exp$ to some (small) starting value. This starting value is the "index," the number of iterations of $\exp$ required is the "level." Thus, in their notation, the level-index number $[3; X]$ would refer to the number $\exp(\exp(\exp(X)))$. Using a system like this, the possibility of overflow is, to all intents and purposes, totally eliminated. Indeed, the highest "level" which could conceivably ever occur in a real-world calculation is 7, so if a computer word were to be split up into level-index form, only three bits would be required for the level, leaving more bits for the index than is the case in the standard floating-point format, where 8 bits or more are used to store the position of the decimal (or binary) point.

A particularly striking feature of level-index arithmetic is that it allows calculations to be made with numbers which are so large that they begin to resemble the fictitious number infinity. For example, the square of the number $[5; 0.87654]$ is precisely the same number $[5; 0.87654]$, correct to all five decimal places in the index. (In their system, all indices are less than unity.)

There is just one drawback with the system as it stands at present. In level-index arithmetic, calculations take about 30 times as long as they do in floating-point form. With computers operating at speeds of millions of instructions per second, this is not so much of a disadvantage as it might first seem, and for some problems the advantages to be gained probably outweigh the loss of speed. But Clenshaw and Olver are currently trying to improve their algorithms to bring the figure down to a factor of 10 or so. At this point it becomes a commercially viable proposition to start designing hardware using the system, and possibly discarding floating-point arithmetic altogether. So, considering the speed of development in the computer field, the day might not be too far away when floating-point overflow is a thing of the past.

35

Add egg to face and take away fame

October 11, 1984

When, earlier this year, the 52-year-old mathematician Louis de Branges solved a notoriously difficult problem first posed by the German mathematician Ludwig Bieberbach in 1916, the mathematical establishment, having largely shunned his work for 30 years, was forced to wipe the egg from their faces and admit that they were wrong, and that de Branges had indeed managed to succeed where so many of them had failed.

The episode illustrates the difficulty faced by professional mathematicians and scientists in such cases: there is always the possibility of the experts being wrong. As the news of de Branges discovery broke, I was reminded of the following report on an unusual piece of mathematical research.

"We have made every effort to understand X's proof. His reasoning is not sufficiently developed for us to judge its correctness, and we can give no idea of it in this report. The author announces that the proposition which is the special object of this memoir is part of a general theory susceptible of many applications. Perhaps it will transpire that the different parts of a theory are mutually clarifying, are easier to grasp together rather than in isolation. We would then suggest that the author publish the whole of the work in order to form a definitive opinion. But, in the state which the part submitted to the Academy now is, we cannot propose to give it approval."

The mathematician was Evariste Galois, and the Academy who issued this put-down in 1831 was the prestigious French Academy of Sciences. Galois is now regarded as one of the greatest mathematicians who ever lived, and the rejected work is correctly recognized as being a work of true genius.

The irony of the situation is that, although mathematics is the only true exact science, with no question of subjective judgements of correctness, the very complexity of the subject means that editors of journals have to decide on the basis of a cursory examination whether or not a piece of submitted work is worth looking at in detail.

Mistakes can be made: highly respected mathematicians can get erroneous work published, while the unknown outsider can encounter great difficulty in getting anyone to look at their work. Although not exactly an unknown outsider, de Branges, of Purdue University in America, was widely regarded as being on the "fringe" of the mathematical establishment, or at least close to it.

In consequence, when de Branges announced his result earlier this year, no one took him seriously. It was only when he visited the University of Leningrad in Russia in May and lectured on his result to a highly skeptical audience of Soviet mathematicians that the realization dawned that he had really succeeded. With the weight of Soviet mathematicians behind it, de Branges work was circulated rapidly around the world, and was recently reported in *Science.*

What is the Bieberbach Conjecture? It concerns infinite polynomials consisting of complex numbers. What it says is that, if the polynomial

$$x + a_2 x^2 + a_3 x^3 + \cdots \quad \text{(ad infinitum)}$$

never has the same value for two values of x of absolute value 1 or less, then the absolute value of the coefficient a_k is at most k for every value of the index k. Though this will seem quite baffling to the average citizen, it can easily be understood by any A-level mathematics student.

The difficulty of the conjecture lies not in its content but in its proof. Over the 70-year period since its formulation, the best that the experts could come up with was the proof that it is true for the first six coefficients, and that all the other coefficients are at most 1.07 times the value of the index in absolute value. In order to get that factor of 1.07 down to 1 as the Bieberbach Conjecture required, de Branges proved a stronger conjecture proposed by the Soviet mathematician Milin in 1971. It was to Milin's group that de Branges lectured in Leningrad. Had Milin not been prepared to listen to him, de Branges' 350-page manuscript would still be gathering dust.

36

Dynastic struggles

November 8, 1984

This week's column is all about numbers which can be chopped off in their prime. A few weeks ago I added a footnote to one of my pieces in which I passed on a question sent to me by a reader. [See Chapter 30.] The response proved somewhat dramatic, and I was deluged with mail. Here is the state of play at the moment.

The original question was this. How many prime numbers are there which remain prime when you chop off the last digit, then the last digit of that number, then the next one, and so on, all the way down to a final one-digit number? For example, 139991 is such a number, since it is a prime and so are the numbers 13999, 1399, 139, 13, and 1. (Professional mathematicians usually exclude 1 from the list of primes, but for the sake of this question it is more fun to include it.)

Many readers were able to find all examples of such numbers. In case you did not try the problem last time, I won't say how to solve it, but it is not too difficult to see how it should be done.

It would be impossible to mention each individual contribution which I received, but one in particular I feel I have to include. It came from Alan Ereira, a London-based historian. His letter was obviously produced using a word-processing package, so I assume he is no stranger to computing. He solved the problem by what he called "a dynastic approach to mathematics." (New readers, there is your clue to the solution.) For all his apologies for not being a mathematician, his work was a perfect piece of mathematical analysis, carried as far as his home micro would allow (without special extra programming), topped off with a conjecture that he had probably found most of them and that there could not be more than 150 or so altogether.

In fact there are exactly 147 such numbers, the largest being 1 979 339 339. (The only other 10-digit example is 1 979 339 333.) If you exclude 1 from the primes, there are exactly 83 such primes, the largest being 73 939 133.

My colleague David Singmaster from the Polytechnic of the South Bank in London recalled seeing an article about this question some years earlier. In 1967, L. M. Chawla, J. E. Maxfield, and A. Muwafi published a paper in the *Journal of Natural Sciences and Mathematics,* Vol. 7 (pages 95–99) in which all of the primes of the kind under discussion are listed. This paper also contains lists of other numbers of a similar nature.

For instance, you could look for primes which remain prime when digits are chopped off either end in any order you like. There are 24 of these if 1 is allowed as a prime, 9 if not. Or you could ask for primes which remain prime when you successively chop primes from the right or when you successively chop primes from the left, but not a mixture of the two choppings. There are 31 of these with 1, 15 without 1.

You could also look for primes which remain prime when you simply chop off digits successively from the left. There are 403 of these less than 10,000 when you regard 1 as a prime, 308 when you exclude 1.

Since this week's column is about curious properties of numbers and their digits, here is a nice little fact I stumbled across in an old puzzle book some months ago. You can arrange the digits 1 to 9 inclusive to form a fraction equal to 1/2, namely 7293/14586. This is vaguely surprising, I think. Truly amazing, you can also find arrangements which give fractions equal to each one of 1/3, 1/4, 1/5, 1/6, 1/7, 1/8, and 1/9. I'll leave you to find all of these. You could use a micro, of course, but I think it would be quicker to figure it out directly.

Finally, what has four processors sharing eight million words of ECL bipolar memory, a 9.5 nanosecond clock time, a 38 nanosecond memory bank cycle time, and a host of other features? The answer is the Cray XMP/48 computer system, claimed to be the most powerful computer system available today.

I pass on this information with the admission that, yes, I too am a computer junkie at heart. Not that I own a micro.* But I do (in secret) enjoy keeping up with the literature produced by the supercomputer firms. Cray Research produces an excellent periodical (strictly for the trade, not for retail) concerning state of the art computing, and it struck me recently how similar in nature it is to many of those much denigrated (by the professionals) micro magazines. Of course, the scale of the applications of the machinery described is different, as is the price of the computers. (The magazine does not quote prices. If you need to know, presumably you cannot afford it.) But at heart, I suspect that there is not a great difference between the home micro "freak" and the billion dollar professional.

---

* Today, in 1993, as I compile this compendium, the Devlin household contains five desktop Macintosh computers, one Macintosh portable computer, a laser printer, a dot-matrix printer, and a *Nintendo* game-playing computer. Such is the march of time.

37

A problem? Hang on while the computer tosses a coin

November 22, 1984

The old adage "When in doubt, toss a coin" is finding increasing application in the world of computers these days. There is something of an irony here, since it is the computer that gives us the means of performing extremely accurate calculations. But, such is the complexity of many of the problems being tackled by computers, that mathematicians are sometimes forced to employ so-called "Monte Carlo methods" to enable the computer to come up with a solution. In such circumstances, in exchange for allowing for the possibility that the computer may fail to come up with a solution or may come up with a wrong answer, the mathematician is able to obtain a speedy solution to an otherwise intractable problem. Then the main question is, just what is the likelihood of something going wrong?

The answer to that question may be highly significant: lives may be at stake. For example, some time ago, mathematicians at SRI in the USA were writing some computer software to operate very efficient aircraft being designed by NASA. Because of the need for split-second decisions, the computer control system had to be able to operate without human intervention. How can you guard against computer failure in such a situation?

The most obvious answer (and the one used by NASA in the Space Shuttle) is to employ several separate computer systems, thereby allowing for one or two to malfunction. But how does the system as a whole decide which computer to believe at any one time? Small discrepancies between the different computers may be due to a significant fault in one of them, or may simply be caused by, say, altimeter readings being taken at slightly different moments. Somehow the system has to be able to make a decision.

It turns out that the mathematical problem which lies behind this situation is one which is extremely difficult to solve within a reasonable time (if at all). But about two years ago, the Israeli mathematician Michael Rabin did produce a very efficient Monte Carlo solution.

When programmed in accordance with Rabin's methods, the multi-computer control system makes its decisions like this. It collects all the information supplied by the different computers, information which it is assumed is not entirely consistent. The system then "tosses a coin" to decide what to do next. That is to say, there are two possible courses of action, which are chosen from quite randomly. Because both of these alternatives are easily dealt with, this takes no time at all in the computer.

The snag is, it is not certain that the result will be a decision (e.g., abort the landing or continue it). All that you can say is that there is a better than even chance that you will reach a sound decision. If you don't get a decision, you try again. And you keep on trying until you do get a decision (or else you crash while you are waiting).

Because the chance of getting no answer each time is less than 1 in 2, if you perform, say, 30 successive tests, the chance of getting no decision is less than 1 in 2^{30}, or about 1 in a billion. So the chance that this procedure will fail you is very small indeed, and you would be unlikely to have to go as far as 30 repetitions, though even that would only take an instant, given present-day computer speeds.

Of course, there will always be that small possibility that you crash before you get a decision: the method is not 100% accurate. But all things being equal, the likelihood of a crash caused by the Monte Carlo aspect of the procedure would seem to be the least of your (or rather the pilot's) worries.

Another of Rabin's Monte Carlo methods can be used to obtain a solution to the so-called "Dining Philosophers Problem," which is related to the organization of computer networks, but I'll postpone discussion of that until another time and give you an example which you can try on your micro. [See Chapter 45.] It is a Monte Carlo method for factorizing large numbers, invented by Pollard some years back.

The obvious approach to adopt in order to find a proper factor of a number N is to look in turn at 2, 3, 5, 7, etc., on up through the primes and see if any of them divide exactly into N. Though this works well for relatively small numbers, for numbers of 50 digits this procedure could take over a billion years, even using the most powerful computer available today. (It is the difficulty of factoring

large numbers which lies behind one of the modern public key cryptographic systems.) For numbers with more than 25 or so digits, you need to adopt more subtle approaches. There are several such, with the best general method available so far succeeding with numbers up to around 60 digits in a couple of hours on a very fast computer.

In 1981, Brent and Pollard were able to factorize a famous 78-digit number using Pollard's Monte Carlo method. (The number concerned is 2 raised to the power 2^8, plus 1, the eighth "Fermat Number.") The method used is this. Pick some number $X(0)$ less than N. Then define a sequence of numbers using the rule

$$X(I+1) = X(I)^2 + 1 \pmod{N}.$$

(i.e., square the last number, add one, and then discard any multiples of N present in the result.) As you proceed, use the extremely efficient Euclidean Algorithm to calculate the highest common factor of the numbers $X(J) - X(I)$ and N, for all J greater than I. The chances are that you will soon discover a highest common factor greater than 1, which will then be a proper factor of N, of course.

The theory behind Pollard's method says that the likelihood of it not giving a factor is small, but not insignificant. (I have simplified the method a bit to produce something you can implement easily at home, but in essence it is all there.) Try it out. For numbers within the capacity of a home micro, you cannot expect any improvement over trial division, but at least you can see how it works.

38

Rabbits do it by numbers

December 6, 1984

Which is the most relevant to the computer scientist, the mouse, the moth or the rabbit?

Everyone knows about mice—small hand-held objects which, when rolled across a desktop, control the display on a computer screen.

And as for moths, they have been immortalized for the computer user in the use of the term "bug" to mean a fault in a computer program. Back in the 1940s, workers at Harvard University built a prototype computer known as the Mark I Automatic Relay Calculator. A malfunction in one of the relays one day was found to be due to the presence of a dead moth. The operator who removed it thereby performed the first officially recorded debugging operation. (Indeed, the remains of the moth were preserved for posterity.)

But for sheer ubiquity throughout computer science, the rabbit easily takes first place. It all goes back to the early 13th century and to the great Italian mathematician Leonardo of Pisa. Being the son of a certain Bonacci, the Latin phrase *Filius Bonacci* (son of Bonacci) gave rise to the name under which he wrote his mathematics text book *Liber Abaci*, namely Fibonacci. In this book, Fibonacci posed the following problem.

A man puts a pair of baby rabbits into an enclosed garden. Assuming that each pair of rabbits in the garden bears a new pair every month, which from the second month on itself becomes productive, how many pairs of rabbits will there be in the garden after one year? Like most mathematics problems, you are supposed to ignore such realistic happenings as death, escape, impotence, or whatever.

It is not hard to see that the number of pairs of rabbits in the garden in each month is given by the numbers in the sequence 1, 1, 2, 3, 5, 8, 13, etc., and that after one year there will be 377 pairs. The general rule which governs the growth of the above sequence is that each number after the second one is equal to the sum of the two previous numbers. This corresponds to the fact that each month, the new rabbit births consists of one pair to each of the newly adult pairs plus one pair for each of the earlier adult pairs.

The sequence of numbers arising from the Fibonacci rabbit problem is called the Fibonacci sequence. It is easily generated on a computer using the rule:

$$U(1) = 1; \quad U(2) = 1;$$
$$U(N + 2) = U(N) + U(N + 1).$$

Mathematicians refer to rules like this, where new numbers are obtained from previously obtained numbers, as "recurrence relations." The Fibonacci sequence arises in many instances in computer science: database structures, sorting techniques, and random number generation, to name three examples.

It could also be argued that the Fibonacci sequence provides a mathematical model of the way in which computer companies grow. Each year the two brightest employees leave to start their own firm, which after a year has grown so big that its two brightest employees .... (Incidentally, if you have not yet read it, try to get hold of Tracy Kidder's book *The Soul of a New Machine* (Penguin), a fascinating account of the everyday lives of simple computer folk on the make.)

The Fibonacci sequence has many surprising properties. For instance, as you proceed along the sequence, the ratios of the successive terms gets closer and closer to the famous "golden ratio" number 1.61803.... And you will obtain the Nth Fibonacci number if you take the golden ratio, raise it to the power N, divide by the square root of 5, and round off the result to the nearest whole number.

Except for 3 (which is $U(4)$), if a Fibonacci number $U(N)$ is prime, then N has to be prime. But the converse is not true. There are prime numbers N such that $U(N)$ is not prime. You might like to try to find examples of such numbers N using your micro. They are not hard to find. It is an open question as to whether there are infinitely many Fibonacci numbers which are prime.

The Fibonacci numbers also arise in nature, in connection with the spiral arrangement of seeds on the face of certain types of sunflower, and with the arrangement of leaves on the stems of several kinds of trees. Getting back to mathematics, it is a pleasant diversion to prove that every positive whole number can be expressed as a sum of Fibonacci numbers, with none used more than once on each occasion.

Finally, and unconnected with Fibonacci numbers (but in the same vein), try the following on your calculator. Enter 2143, divide by 22, then press the square root button twice in succession. What do you notice about the answer?

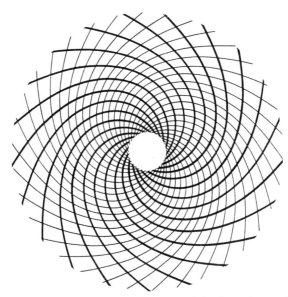

FIGURE 13.   Geometry in nature. The complex, spiral pattern of the florets in a sunflower. In addition to the geometry, nature also exhibits some elementary number theory: the number of spirals are adjacent Fibonacci numbers: there are 21 clockwise and 34 anticlockwise spirals.

39

Add mission

December 13, 1984

A Paris-based Japanese mathematician called H. Matsumoto has claimed to have solved the most famous open problem in pure mathematics—the Riemann Hypothesis, first raised in 1859 by the German mathematician Bernhard Riemann. Matsumoto made his startling and totally unexpected claim at a mathematics meeting in Paris in November, but such is the enormity of the claim that mathematicians were at first reluctant to admit even the possibility that Matsumoto had succeeded where so many had failed in the past. Even now, no one is prepared to say with certainty that the proof is correct, but those who have read it admit that they have been unable to find any mistakes.

If the result is true, and no one doubts Matsumoto's ability to obtain such a discovery, it will have far-reaching consequences throughout large parts of current mathematics. The Riemann Hypothesis is closely connected with the distribution of the prime numbers amongst all whole numbers, which fact alone suggests its fundamental nature, but its effect spreads much further than that. The list of eminent mathematicians who have worked on the problem reads like a Who's Who of Nineteenth and Twentieth Century Mathematicians, and most present-day mathematicians had long given up hope of ever seeing it proved.

The hypothesis asserts that all of the solutions (amongst complex numbers) of a certain equation have a certain form. Since there are known to be infinitely many solutions, there was no chance of proving the hypothesis by direct computation of all the solutions.

But there was always the possibility that a computer search might produce by chance a solution not of the desired form, thereby disproving the hypothesis. Using modern computers, mathematicians have calculated over 3,000 million solutions, all of which turned out to have the stated form, a fact which tended to confirm the hypothesis. Many highly significant mathematical discoveries were made in the course of attempts to prove the hypothesis by rigorous mathematical means.

It is likely that all mathematicians would agree that the Riemann Hypothesis is (or was) the biggest open problem in mathematics. Indeed, the famous mathematician David Hilbert included it as the fourth item in a list of the most important problems facing mathematics in the 20th century, given at the Second International Congress of Mathematics held in Paris in 1900. (The Riemann Hypothesis is the only problem in Hilbert's list that has not been disposed of until now.) Such is the highly technical nature of the Riemann Hypothesis, and the claimed proof, that the majority of mathematicians are unlikely to be in a position to follow Matsumoto's reasoning, which involves the application of methods of modern functional analysis to structures called Adele groups, and which follows suggestions made by the mathematician André Weil some 20 years ago.

However, there can be few mathematicians whose work will not be in some way affected by such a result. I shall give further details about the Riemann Hypothesis in a few weeks time. [See Chapter 41.]

Postscript.   The alleged proof turned out to be false.

40

Chimney sweep for Santa

December 20, 1984

One day shortly before Christmas, Santa's assistant, the faithful gnome Graudnia, came home to find his master looking very glum.

"What's up, Santa?" asked Graudnia.

"It's no good," replied Santa sadly, "there are so many houses to visit on Christmas Eve that I cannot possibly get to them all in the time available. I've been trying for days to work out a route to follow, but it just seems impossible. Even with just 100 houses to visit, the number of possible routings between them is a number with 158 digits. And I have many more than 100 houses to visit."

"It seems to me that what you need is a computer," said Graudnia. "Why don't you talk to your neighbor Jack Snowfield,* he knows all about that kind of thing."

So Santa did just that.

"Well, Santa," said Mr. Snowfield, knocking the ash from his pipe, "as you know I usually recommend those tiny lap held portable computers, but for your problem you have so much data to handle that I think you will need something really big, like one of those giant Cray supercomputers which can perform billions of operations per second." With which he left to try to send himself a Christmas message by way of computer links through San Francisco and Tokyo.

A few days later, Santa was admiring his brand new supercomputer. (This is a fairy story remember. We will also have to pretend that the system works the first time!)

"Well," said Graudnia, "there it is, but how do we program it to solve your routing problem?"

"Simple," replied Santa, "I've asked our old friend Mr. Hacker, the computer programmer, to come over and help us. He should be here any minute now."

Just then there was a ring at the door, and Graudnia opened it to find Mr. Hacker standing there.

"Well, Santa," said Mr. Hacker, sipping his cocoa some time later. "I've asked around and it seems that your problem is well known. It is usually called the Travelling Salesman Problem, or TSP for short. It was first studied in 1930 by a Viennese mathematician called Karl Menger. The general problem is that you have a number of houses or cities to visit, you know the distances between each pair of them, and you have to work out the shortest route which takes you to each of them in turn.

"The most obvious approach is to look at all possible routes, work out the distance travelled for each, and then pick the shortest. But with just 10 cities to visit there are 3,628,800

---

* The editor of the *Guardian* computer section was Jack Schofield.

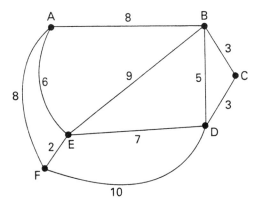

FIGURE 14.   The Travelling Salesman Problem. What is the shortest route that enables the salesman to visit each of the cities A through F?

possible routes to look at, and with 25 cities a staggering 16 followed by 25 noughts routes. This number is bigger than the age of the universe measured in microseconds, so even your supercomputer cannot handle it that way."

"But wait a minute," said Santa, "even I can see that you don't have to look at every possible route. It makes no sense looking at routes which keep jumping huge distances each time. Why don't we just look at routes which try to travel the shortest distance each time. Surely that will work?"

"Yes, that's what I thought at first," replied Mr. Hacker, "but it seems that mathematicians have shown that this does not always work. In fact the situation is a lot worse than you might think. There is a whole collection of computation problems which mathematicians have studied called NP problems. I'm not sure what the letters NP stand for here, but the NP problems include many important problems to do with costing, scheduling networking, computer chip designing, coding and the like, problems which mathematicians have tried very hard to solve without any success."

"And I suppose you are going to say that my TSP is in this collection NP," said Santa, more gloomy than ever now.

"Worse, I'm afraid," replied Mr. Hacker. "What mathematicians have been able to show is that if the TSP could be solved, then every single problem in the collection NP could be solved. To put it another way, if someone could write a computer program which solves the TSP, it would be a simple matter to modify it to solve any of the other NP problems. So you see, your problem is as hard as, if not harder, than all of the problems in NP which no one has been able to solve. To put it bluntly, most of the experts I talked to think the TSP simply does not have a workable solution."

"So that is that," sighed Santa. "For the first time ever I will not be able to deliver all my presents."

"Not quite," said Mr. Hacker, "I have been saving the good news until last. Although no one expects that there can be a workable method which provides an exact solution to the TSP, there are some good methods which will give you a route which is very nearly optimal, say within 10% of the shortest route. For most applications, this ought to be acceptable, and I'm sure it will be adequate for your purposes. And mathematicians are working all the time to improve such approximation algorithms, as they are called.

"For example, only recently a young Indian mathematician called Narendra Karmarkar working at Bell Laboratories in the United States discovered a new method for solving the Linear Programming Problem, a problem which in some ways resembles the TSP. Apparently this method works much better than the previous one, which was called the Simplex Method. So you see there is no need to be gloomy, is there?"

- *Time*'s Man of the Year:

  Deng Xiaopeng

- Starvation in Africa: Ethiopia to Mali

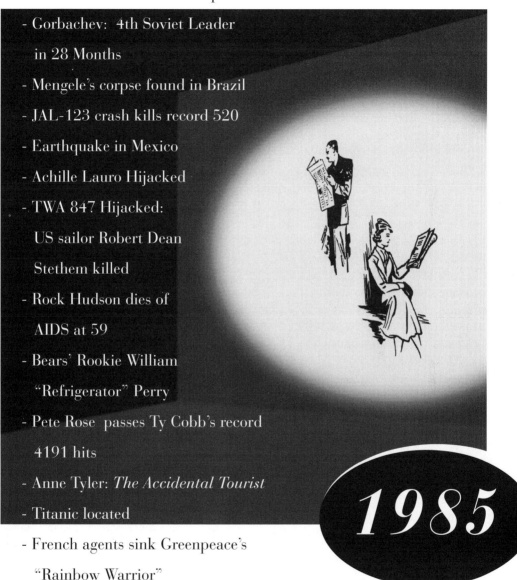

- Gorbachev: 4th Soviet Leader

  in 28 Months

- Mengele's corpse found in Brazil

- JAL-123 crash kills record 520

- Earthquake in Mexico

- Achille Lauro Hijacked

- TWA 847 Hijacked:

  US sailor Robert Dean

  Stethem killed

- Rock Hudson dies of

  AIDS at 59

- Bears' Rookie William

  "Refrigerator" Perry

- Pete Rose passes Ty Cobb's record

  4191 hits

- Anne Tyler: *The Accidental Tourist*

- Titanic located

- French agents sink Greenpeace's

  "Rainbow Warrior"

1985

41

Has the last great math mystery been unravelled?
January 3, 1985

Ask any professional mathematician what is the greatest unsolved problem in mathematics and you are virtually certain to be told: "The Riemann Hypothesis." First formulated in 1859 by the German mathematician Bernhard Riemann, this problem has resisted attempts to prove (or disprove) it by many of the world's most distinguished mathematicians, and few present-day mathematicians expect to see it resolved within their lifetime. But now it seems that it may—just may—have been solved. On November 9 of last year, the Paris-based Japanese mathematician H. Matsumoto announced that he had proved the hypothesis.

The story begins back in the eighteenth century with attempts by the German mathematician Leonhard Euler to establish what is now known as the Prime Number Theorem.

A few moments with paper and pencil (or its modern equivalent, several hundred pounds worth of microcomputer) are enough to convince you that, as you proceed up through the positive whole numbers 1, 2, 3, etc., the occurrence of prime numbers (numbers having no exact divisors other than themselves and 1) becomes less frequent. On a fairly local scale, there seems to be no obvious pattern to the way the primes thin out: you keep finding long stretches without any primes and then runs of several primes close together, in a seemingly haphazard fashion.

But, if you step back and take a more global view, some kind of pattern does seem to emerge. It was probably Gauss who first noticed that if $P(n)$ is used to denote the number of primes less than n, then the larger n becomes, the closer the 'density' function $P(n)/n$ gets to the value of $l/\log(n)$.

This remarkable observation, based solely on numerical evidence in the first instance, indicates that there is a hidden connection between the prime numbers, positive, whole quantities, and the natural logarithm function, which has to do with infinitely small subdivisions between numbers. To understand the behavior of the prime numbers it appears you have to look at numbers closely tied up with the infinite. The first rigorous mathematical proof of Gauss's observation (the Prime Number Theorem) was provided by Hadamard and de la Vallée Poussin in 1896. Neither their proofs nor any of the subsequent ones are accessible to anyone but the specialist in number theory.

To return to Euler, some time around 1740 he introduced the "Zeta function," which is defined for real numbers s greater than 1 as the sum of the infinite addition

$$\text{Zeta}(s) = 1 + \frac{1}{2^s} + \frac{1}{3^s} + \cdots.$$

Euler showed that this strange looking function is closely connected with the prime numbers, and thus provides a definite link between the primes and the infinite.

In a paper written in 1859 (in German), entitled "On The Number Of Primes Less Than A Given Magnitude" (see *The Collected Works of Bernhard Riemann,* Dover Publications), Riemann took matters a stage further regarding the Zeta function, by extending it from the real numbers greater than 1 to be defined on all complex numbers (numbers involving the square root of minus one). To the man in the street, the very idea that something as concrete as the prime numbers can be connected with square roots of negative quantities may well seem incredible, but Riemann's move proved decisive, and present-day mathematicians are well aware of the intimate connection between all kinds of numbers, "concrete" or "imaginary." The Zeta function is nowadays known as the "Riemann Zeta Function." Entire books have been written about this one function (for example, *Riemann's Zeta Function* by H. M. Edwards, Academic Press, 1974).

The Zeta function is connected with the prime numbers in many ways. One connection involves the solutions to the equation

$$\text{Zeta}(s) = 0.$$

Riemann observed that all the solutions he calculated had the same form, namely they were all complex numbers obtained by adding exactly 1/2 to the square root of a negative number. He put forward the hypothesis that all solutions would have this form: this is the "Riemann Hypothesis." When it proved impossible to prove that the hypothesis was true, mathematicians took to calculating solutions to the Riemann equation in the hope that they might find one not of the stated form, and thereby disprove the conjecture. Gram calculated the first 15, and more were found by Backlund and then Hutchinson, until Titchmarsh and Comrie brought the total to 1,041

solutions in 1936. All were of the form predicted by Riemann.

Then came the computer. Before the first computer had even been built, the great British computer pioneer Alan Turing had developed methods for dealing with the problem suitable for running on one, and by Lehmer's work in 1956, some 250,000 solutions had been found. The most recent computations, in 1982, by Van de Lune and te Riele, took the total number of calculated solutions to 300 million and one. Again, all have the form predicted by Riemann.

Overwhelming evidence you might think! Unfortunately, there are infinitely many solutions to the Riemann equation, and it can be argued that if the equation were to have a solution of a different kind such a solution would perforce be beyond the range of any conceivable computer. And a conjecture very closely connected with the Riemann Hypothesis, called the Merten's Conjecture, was proved false early in 1983 despite having been computer verified for 10 billion cases! [See Chapter 19.]

But Matsumoto's result, if correct, will show that, in the case of the Riemann Hypothesis, the computer evidence is not at variance with the fact. The approach adopted by Matsumoto follows suggestions made by André Weil some 20 years ago, involving the application of methods of modern functional analysis to structures known as Adele groups. Though it is unlikely that the majority of mathematicians will be able to follow the purported proof without a great deal of effort, if it turns out to be correct many of them will be affected by the result in one way or another. It may also have an affect on society at large. Modern developments in cryptography have meant that results about prime numbers might result in (at least the possibility of) the electronic theft of our bank accounts! And the Riemann Hypothesis is no ordinary result about prime numbers! You may be about to see some pretty rapid events in the near future.

Postscript.   The claimed proof was found to be incorrect. At the time of assembling this compilation, in 1993, the Riemann Hypothesis remains unproven.

42

The short-cut solution

January 17, 1985

Possibly the greatest challenge facing modern man is that of securing the best deal when booking an airline flight. Trying to minimize your outlay while at the same time satisfying the various constraints imposed by your schedule and by the different requirements accompanying the numerous special offers seems to require a Ph.D. in mathematics at the very least. So, as one whose infrequent forays to exotic places tend to be supported by research allowances which appear to be calculated on the basis of travel by bicycle, I gave more than a passing glance to the recent announcement that a young Indian mathematician working at Bell Laboratories had discovered an exciting new method of solving this kind of problem.

The Linear Programming Problem has been around as long as anyone can remember, and arises in all walks of life, particularly in economics, industry, engineering, transport, and defense. You have something you want to optimize (e.g., make your costs as low as possible, maximize your profits, increase your chance of survival in some risky venture, etc.), and you can achieve this by altering any one or more of a number of parameters, these parameters being subject to various constraints.

For example suppose you own a factory which makes both widgets and wodgets, with the current market price of the widget being twice that of the wodget. The manufacture of widgets and wodgets require some common raw materials and some unique to each, and your highly skilled workforce can only produce one kind of product at any one time.

Of course you wish to maximize your profit. Ideally you work out which of your two products brings in the greatest profit and just concentrate on that. But life is never so simple. You find that by switching from one to the other every now and then you can maintain your machinery virtually without shutting down any part of your plant for service. Then there is the problem of supply and storage of the raw materials you require, as well as the storage of the finished products (which, you will realize, involve totally different conditions). And your sales people tell you that there are constraints on just how many widgets the market can stand without a ready supply of wodgets to go with them. To say nothing of potential union problems if you don't tread warily. How on earth do you figure out how to maximize your profit while satisfying all of these constraints?

Now imagine a genuine situation which a modern businessman faces every day, where the number of constraints is vastly greater than in the above example. The sheer number of different permutations of the available options prevents you from adopting the most obvious approach of calculating your profit for each possible combination and then simply picking out the best. But provided that all of your constraints are "linear" (in the case of two parameters, that means that if you draw a graph of any one of your constraints it will be a straight line), there is a reasonably efficient way of performing the calculation. (The use of the word "linear" in Linear Programming stems from this requirement. Most real life optimization problems either already are, or else can be reformulated as, linear programming problems in this sense.)

The method is known as the *simplex method*, and was invented by the American mathematician George Dantzig (now at Stanford University) in 1947. Until recently, this method was by far the most successful way of solving linear programming problems, and the major computer manufacturers all supply commercially written programs for performing Dantzig's simplex algorithm.

The idea behind the simplex algorithm is this. You first think of the problem in geometric terms. To take the simplest case first, if the problem only involves two parameters, say x and y, then since the linear equations in two variables represent straight lines on an ordinary graph, the various constraints of

your problem correspond to straight lines on a graph.

Now, if you draw a number of straight lines on a piece of paper they will trace out a polygon; a triangle if there are three lines, a quadrilateral with four, a pentagon with five, and so on. The polygon you obtain in this way is a geometric representation of your two-parameter problem constraints.

With three parameters, you get a three-dimensional geometric realization of the constraints, namely a polyhedron (e.g., a tetrahedron, a five-sided "box," or whatever).

For greater numbers of parameters, the geometric realization is, of necessity, purely abstract, since more than three dimensions will be necessary: if the problem has N parameters, the geometric figure corresponding to its constraints will be what is called an N-dimensional "polytope." (Incidentally, this indicates how a mathematical concept as bizarre as a 100-dimensional "polytope" can be of real use to the hard-pressed business executive!)

The key to the Simplex Method is the fact that the values of the parameters which give you the optimum solution you require will be the coordinates of one of the vertices (corners) of this polytope. The aim of the simplex algorithm is to find this particular vertex.

It does this by starting at one particular vertex (the closer this vertex is to the

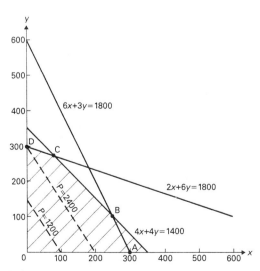

FIGURE 15.   **The Simplex Method.** A useful way to turn an optimization problem into a geometry question.

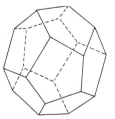
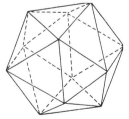

FIGURE 16. Polytopes. These familiar polygons are three-dimensional examples of the "polytopes" that arise in optimization problems. Real-life optimization problems usually give rise to polytopes in many dimensions.

final optimal vertex the better) and then proceeding from vertex to adjacent vertex until the optimal one is found. In the case of a two-parameter problem this is easy, since there is only the choice of "clockwise" or "anticlockwise" to consider, but it is obvious that even a moderately complicated three-dimensional polyhedron offers numerous branching choices of route. The simplex algorithm works by making the best choice of route available at each step. (Remember, until it finds it, the algorithm has no idea where the optimal point is, so it cannot just "aim straight at it.")

It is possible to construct artificial problems which result in this approach taking a prohibitively long time to find the optimal vertex, but for most real life problems it seems to work well. And though in the early days of computers it was customary to take a short holiday while the program worked away on a problem with only a few hundred parameters, an efficiently written Assembly Language simplex program running on a fast mainframe computer can nowadays handle a typical 1,000 parameter problem in about five minutes.

But for "real time" tasks, such as the control of aircraft or nuclear power plants, even this is far too long, so mathematicians have continued to look for a better method. In 1979, the Russian mathematician L. G. Khachiyan discovered an alternative method which was theoretically faster than the simplex method, called the *ellipsoid method,* but in practice it performed much slower than the simplex method, the advantage only showing up on the artificial examples constructed to beat simplex.

Then, last year, Dr. Narendra Karmarkar, the son of a mathematician, who grew up in Poona and studied for his doctorate at Berkeley, California, before joining Bell Laboratories in New Jersey in 1983, discovered a new method of tackling the problem. Tested on a typical 5,000 parameter problem (there is a version of Murphy's Law that says that the "typical" problems which arise are always at the limits of current computational ability), Karmarkar's algorithm found the solution 50 times faster than a good commercially produced simplex program (the MPSX/370). Since this latter product is written in Assembly Language and utilizes the latest methods of pipelined computation, whereas Karmarkar's program was just a straightforward Fortran program, it can be inferred that the new method is significantly faster than simplex, and will rapidly supersede it, perhaps even making "real time" control problems feasible.

The key mathematical idea behind Karmarkar's algorithm is intuitively an obvious one. To go from an initial vertex of the constraint polytope to the optimal one it would be quicker to take a short cut across the interior of the polytope, rather than meander along the edges. The problem is that if you try to do this, once you lose contact with the outer surface, then, like an astronaut floating in space without a lifeline, you can easily go off course. (The more so since you don't know exactly where you should be heading.)

Karmarkar overcomes this difficulty by proceeding in relatively small steps; after each one you perform a mathematical deformation of the interior of the polytope, with the result that in the deformed geometry there is an "obvious" direction to proceed, and you then move in the direction in the original space which corresponds to this deformed direction under the chosen deformation. In "no time at all" you arrive at the optimal vertex.

Intuitively, the procedure can be thought of like the flight of a guided missile, which constantly alters its direction until it homes in on the target. One of the most depressing thoughts is that this will probably be one of the first applications of the new method.

43

The software jungle
January 31, 1985

In many ways the introduction of micros into schools was similar to the decision to send the Task Force to the Falklands. There were powerful arguments in favor of both enterprises, but in each case the final decision was taken hastily, with the inevitable pressure brought about by that most unstoppable factor, the fragile egos of Government ministers. This is not the place to resurrect the Falklands issue. But as far as micros into schools is concerned, the haste with which it was carried out created problems which could have been avoided had more thought gone into the issue.

As the experts said at the time (to no avail), there is no point in simply supplying the hardware. Without good software it can be positively harmful. And good software is difficult to produce. With the best will in the world (and heaven knows there is plenty of that about in the teaching profession), it cannot be produced by paying teachers to do it at home in their spare time. To do the job properly (and if it is not done properly it should not be done at all), you need a great deal of expertise, involving educationalists, mathematicians (if the software is to be at all mathematical), and professional programmers (if the technology is to be used to its proper potential). In short, good software is expensive. Very expensive.

Similar remarks can be made about books, of course. But there are two significant differences. First, books have been with us for a very long time, and there is a wealth of accumulated wisdom when it comes to writing them. Second, and this is where the danger lies, the great majority of the users of computer technology in the schools are not at all familiar with it, with the result that there is very little chance that inherently bad programs will be seen for what they are. No teacher would choose a book simply because it has a nice glossy cover, but how many teachers are able to properly evaluate a computer program?

I'll admit that I, for one, do not feel qualified to judge an educational program unaided. Certainly I could comment on any mathematical content—if the software supplier were to supply me with details. But as far as any other aspect goes, I simply do not know enough about the subject(s). All I can go on is the surface gloss, and this can be produced in copious amounts on a modern micro.

I'm not trying to attack teachers. Nor the brave souls who struggled to produce some programs to run on the suddenly acquired school machines (I even did a bit of this myself). The main culprit, as usual, is a Government which seems singularly inept when it come to anything to do with education. Education does not come cheap. Particularly so when computers are involved. The poor struggling teacher has to pick up the pieces and make the best of what is available. In the meantime, my advice to teachers would be to tread warily.

I spoke about danger. Was I overreacting? I don't know. And what is more, I doubt if anyone does. We simply do not have enough experience yet. What I do know (because I have been informed by experts whose judgment I respect) is that there are various programs currently in use in schools up and down the country which do not stand up to a proper analysis.

Notable in this class are those programs which profess to develop the ability of the student to make balanced decisions. (You know the kind of thing. A list of options is given, you choose one, another list comes up, and so on, and at the end you get an evaluation of your decisions. There are lots of variations on this theme.)

To be of any real educational value, such a program should provide a good simulation of a real life decision-making procedure. (Unless it is regarded as useful for the student simply to figure out just what the writer of the program had in mind.) Unfortunately, this area of computer science, namely Artificial Intelligence, is known to present great problems, problems which have not been satisfactorily solved so far, and which are not likely to be solved in a manner which can be implemented on your average school micro. The potential danger is that a whole generation will grow up with the impression that the computer can help them to make valued decisions, which (except in a highly restricted sense) is certainly not the case.

No doubt things will gradually improve. (Indeed there are some "educational" programs on the market which strike me as very good indeed.) But it would have been far better had the whole enterprise been properly thought out in the first place. In the meantime, my advice to teachers would be to beware. No, I am not claiming to know all there is on this subject. I know just about enough to realize how little I really do know. And that worries me.

44

Measured smile

February 14, 1985

Once I had got beyond high school, I thought math was fun, and I still do. And judging by the mail I receive, so do many readers of this page. But to some people it would appear that like religion and royalty, math is far too serious to be treated in a light-hearted fashion. Occasionally I get letters taking me to task for not giving the subject the "gravity" it deserves.

So this week's piece is written especially for fun-loving mathematicians. Though I cannot reply to everyone who writes to me in person, I would like to hear from anyone who makes progress on any of the problems.

A whole number is called *digitally divisible,* if none of its digits is zero and it is divisible by the sum of its digits. For example, 322 has this property; the sum of its digits is 7, and 7 does divide into 322 (without remainder). It is known that there are infinitely many such numbers.

In particular, for any value of N, the number which consists of $3N$ successive 1's in decimal representation has this property. (The proof of this fact is a nice exercise.) But how easy is it to find such numbers? This is obviously a problem for which a micro is ideally suited. Rather than try it myself, I shall sit back and leave the hard work to you.

The next problem extends one I have given before on this page. It concerns making fractions using the digits 1 to 9. You can arrange the digits 1 to 9 into a fraction to obtain, af-

ter cancelling, each one of the numbers 1/2 to 1/9. For example, 1/2 = 7293/14586. Without a zero, you cannot obtain 1/10 of course, but what about 1/11, 1/12, and so on? Some of the fractions have more than one such representation. For instance, 1/2 can also be written as 6792/13584 (and in other ways). When you have written a program which finds such solutions, modify it to calculate the number of different solutions for each of 1/2, 1/3, and so on, up through 1/9, 1/11, etc., as far as possible (excluding the multiples of 1/10, of course). Does a pattern emerge? If so, can you think of any explanation?

My last problem is an old classic, dating back at least to the 1930s, and it has a history of gobbling up hours of computer time even back in the days when this was an expensive commodity. It is extremely easy to state. Think of a number. If it is even, halve it. If it is odd, multiply it by 3 and then add 1. Now do the same to the answer, then to that answer, and so on.

For example, suppose you start with 1. This is odd, so multiply by 3 and add 1, which gives you 4. Since 4 is even, you now divide by 2, to give 2. Since 2 is even, you divide it by 2 and get 1. At this point, you are back where you started, of course, and from now on the process repeats itself indefinitely.

Now try starting with 3. This gives the numbers 10, 5, 16, 8, 4, 2, 1, 4, 2, 1, and so on. So after an initial trip up to 16, you tumble down to the 1, 4, 2 loop.

This is where the story gets interesting. No matter what number you start with, this process always seems to lead you to the 1, 4, 2 loop. Nabuo Yoneda of the University of Tokyo has used a computer to test all starting values up to a million million. Every one leads to the 1, 4, 2 loop. No one knows if there is a mathematical reason for this, or whether it is just a fluke which fails for extremely large numbers.

Yoneda's result probably means that the home micro user has little chance of breaking new ground, but it is interesting to examine the lower slopes yourself. In BASIC, the essential part of a program to investigate the problem could be written like this:

$$\text{if } N \bmod 2 = 0 \text{ then } N = N/2$$

$$\text{else } N = 3 * N + 1$$

Once you have a program which produces the results of the procedure for a given starting value, add clauses to count the number of steps before you reach 1 and to record the highest value reached. (With starting value 27 you get up to 9,232 before coming down to 1 after a total of 111 steps.) Or you could arrange for the values of the sequence to be plotted as a graph of sequence number against iteration number. For the starting value of 27 mentioned a moment ago, and for many others, you get a graph which rises and falls many times, a picture which has led some writers to refer to the procedure as generating "hailstone numbers," reminiscent of the way hailstones form in the updrafts and downdrafts of clouds. You are unlikely to discover anything new about hailstone numbers, but you should get a lot of enjoyment from watching them live through the ups and downs of their youth before finishing up in their inevitable treadmill maturity.

45

Food for thought

February 28, 1985

Increasingly, computers are being connected together to form networks. This creates problems which can be very difficult to solve for the mathematician. A classic example is the Dining Philosophers Problem, formulated by the Dutch computer scientist Edwin Dijkstra.

Imagine a group of philosophers sitting round a table at dinner time. In the center of the table is a large plateful of spaghetti. Between each pair of philosophers is a fork. Most of the time the philosophers are talking and thinking (as philosophers do), but from time to time one or more of them feels hungry and wants to take a helping of spaghetti. Now, it takes two forks to serve yourself from the bowl, so the hungry philosopher needs to pick up both of the forks adjacent to him. If either of these two forks is being used, it is impossible to get any spaghetti at that moment.

Now suppose that at some stage in the proceedings, all the philosophers want to eat at the same time. Being rational beings, capable of performing only one action at a time in a serial fashion (philosophers get like that after a time!), they each reach first of all for the fork on their right and pick it up. So far so good. Now they turn to their left for the second fork, and what do they find? Someone else has got there first. Stalemate.

Being terribly egalitarian and rational, each one waits patiently for the second fork to become available. Which never occurs, of course. The result is that the entire group of philosophers slowly starves to death.

Ignoring for the moment the rumor that this entire scenario was carefully considered by the Cabinet following the rejection by Oxford University of an honorary degree for Mrs. Thatcher a few weeks ago, what has Dijkstra's problem got to do with computer networks?

Suppose that, instead of philosophers sitting round a table, you have a collection of computers, connected together in the form of a ring with each one connected to its two neighbors. Most of the time each computer gets on quietly with its own task, but occasionally it is necessary for one to communicate with one or both of its neighbors.

The problem arises, how can you program each computer so as to avoid a stalemate situation as with the starving philosophers? Whatever you may think of philosophers, they are, by and large, more enterprising than your average computer, which will happily spend the rest of its life waiting for an event which will never happen, unless the careful programmer foresees that such a situation might one day arise and guards against it.

One obvious solution would be to designate one computer the "boss" and let it make all decisions about when two others can communicate. This solution is often adopted in the construction of networks. But what if, as with the philosophers, each computer is to be on equal terms with all the others? The problem of avoiding deadlock is now an extremely difficult one, and was only solved fairly recently.

The solution found by Michael Rabin and Daniel Lehmann of the Hebrew University of Jerusalem is an example of what is known as a Monte Carlo method, which depends for its success on probability theory. Imagine the philosophers again.

When one of them feels hungry, he randomly chooses between the left fork and the right one. (This has to be a truly random choice; if there is any rhyme or reason behind the choice the method will not succeed in avoiding a stalemate.) Having chosen a fork, say the right one, the philosopher waits for this one to become available. (He could continue talking and thinking, of course, so this waiting time need not be lost.)

As soon as the right fork becomes available, the philosopher picks it up and then looks for the other one, the left one in this case. If it is available, he picks it up and eats.

If it is not, he immediately puts down his existing fork and starts the whole process again from the beginning, once more making a random choice of direction.

Looked at from the point of view of the individual philosopher, starting again when you seem to be half-way there looks like a retrograde step, of course, but from the point of view of the entire group this is by far the best way to proceed. For Rabin and Lehmann have proved that, if this procedure is followed, a stalemate will not arise. By extending the procedure, they can also show that no one will starve, i.e., provided they wait long enough, each person will eventually get to eat. The mathematics involved here is by no means trivial, though the procedure itself is relatively simple to implement.

The Monte Carlo solution to the Dining Philosophers Problem is just one example of an increasing use of statistical methods in computer science and mathematics, and to some extent is a consequence of the increasing complexity of computers and the uses to which they are put.

46

Playing the negadecimal game

March 14, 1985

In previous installments of this column I have talked about the use of different number bases for representing numbers. [See, for example, Chapters 5 and 32.] The most familiar system to us is, of course, the decimal system. This makes use of the ten digits 0, 1, 2, 3, 4, 5, 6, 7, 8, 9 to represent numbers, and in our arithmetic we require a units column, a tens column, a hundreds column, and so on.

Computers make use of the binary system, where there are just two binary digits (or bits) 0 and 1, and where arithmetic requires a units column, a twos column, a fours column, an eights column, and so on. In decimal arithmetic we must "carry" whenever a multiple of 10 occurs in any column, in binary whenever a multiple of 2 occurs.

Mankind makes use of decimal notation because people have ten fingers; computers use binary notation because a computer is, at heart, a two-state machine, the current in a circuit being either on or off.

The main problem with the binary system as far as humans are concerned is that it takes impossibly long "words" to denote even moderately large numbers. For instance, the number 229 expressed in binary notation is 11100101. Starting from the right, i.e., from the units, this number is:

$$(1 \times 1) + (0 \times 2) + (1 \times 4) + (0 \times 8) + (0 \times 16)$$
$$+ (1 \times 32) + (1 \times 64) + (1 \times 128) = 229.$$

Consequently, all modern computers have built-in routines which automatically convert numbers from decimal form to binary and back again, allowing human operators to communicate with the machine in everyday decimal form.

But sometimes it is necessary for the programmer to handle the numbers in the machine in the form (i.e., binary) that they are stored in the memory. This can be made easier by utilizing the so-called hexadecimal system, that is the number system with base 16.

This has the effect of replacing four columns of binary by just one column of hexadecimal (because 16 is the size of the fifth column in binary, starting from the right). In other words, every hexadecimal digit specified by the programmer determines four binary digits in the computer. There are fifteen hexadecimal (or "hex") digits:

$$0, 1, 2, 3, 4, 5, 6, 7, 8, 9, A, B, C, D, E, F.$$

So, for example, the hexadecimal number 1BF5 represents the number:

$$(5 \times 1) + (15 \times 16) + (11 \times 256) + (1 \times 4096)$$

i.e., 7157 in decimal notation. (Because 256 = 16 × 16 and 4096 = 16 × 16 × 16.) In practice, programmers usually need to make use only of hexadecimal numbers involving two hex digits. This is because the bits that make up a computer word are grouped into bytes, collections of eight consecutive bits. Two hex digits completely specify all the bits in one byte.

So far, I have only been talking about positive whole numbers. Fractions can be handled by specifying a decimal/binary/hexadecimal point. There are various ways of doing this, but that is another story. What I want to look at now is how negative numbers are handled. The most common method used in computers is for one bit of each computer word to be reserved to denote the sign of the number (say with a 0 denoting a positive number, a 1 a negative number).

Calculator displays usually work like this, except that on the display a minus sign appears at the left-hand location instead of a 1, and nothing appears when the number is positive. The computer hardware is then constructed to keep track of the signs of numbers during arithmetic operations. But other methods have been considered.

Anyone who has used one of those mechanical calculating machines that used to fill offices 20 years ago will appreciate one of these methods. On those ancient machines (which worked in decimal arithmetic), if you subtracted 1 from 0, the machine would display an entire row of 9s. This is because, as far as the machine was concerned, this really was -1. If you added 1 to a full string of 9s, you would get a carry all the way along the number, and off the left-hand end, leaving zero; i.e., $(-1) + 1 = 0$. Similarly, in a computer, a complete row of 1s could be used to represent -1. In both these systems, it is easy to see how any negative number could be represented, not just -1.

Even more intriguing is to use a negative base in the number representation. For instance, you could represent your numbers in the negadecimal system, where the base is -10. In this system, the number 211 is equal to the decimal number

$$1 + (1 \times -10) + (2 \times -10 \times -10),$$

i.e., 191. Again, 35 in negadecimal is the same as the decimal number

$$5 + (3 \times -10) = -25.$$

So in negadecimal, negative numbers (like -25) can be represented without negative signs being necessary. In fact, any number can be written out in negadecimal notation, and regardless of whether the decimal number is positive or negative, no sign is necessary in negadecimal.

It is quite amusing to spend a few minutes converting numbers from decimal to negadecimal and back again, and to work out how to perform addition and multiplication of numbers written in negadecimal notation. An ordinary microcomputer can easily be programmed to act as a calculator for numbers in negadecimal format, and this makes a nice exercise in computer programming.

The negabinary system could be used as a basis for computer hardware design, and this would mean that it would not be necessary to have a sign bit in computer words. Though this has been seriously considered, I am not aware that it was ever used in practice. [See Chapter 51.]

47

The taxi cab that caused a conundrum

March 28, 1985

What does the number 1729 mean to you? Anything special? While you are pondering on that, let me say a few words about the Indian mathematician Srinivasa Ramanujan, to whom 1729 had considerable significance. Born in Erode in southern India, in 1887, to a fairly poor family, Ramanujan showed at an early age that he had a particular talent for mathematics. Before he was ten he had read large parts of an English university mathematics textbook he had access to, and was able to solve many of the problems in the book. He could also perform amazing arithmetical calculations in his head, and was able to memorize the value of π to many decimal places.

Unfortunately, by the time he got to university, his interest in mathematics had become so all-consuming that he neglected his other subjects and became a drop-out. Several attempts to regain a university place failed, so in 1912, when he was 25 years old, he became a clerk in the office at the Port Trust of Madras at a salary of £30 a year.

Fortunately for mathematics (as well as for Ramanujan himself), the Cambridge mathematician G. H. Hardy got to know of some of Ramanujan's work. Though the authorities at the Indian universities were not able to recognize a genius when they had one, Hardy was, and so by 1914 he had brought Ramanujan to Cambridge. Despite the fact that Ramanujan never did receive a proper education in

mathematics, and to his death was ignorant of many of the most basic mathematical concepts of the time, his genius was such that, partly in collaboration with other Cambridge mathematicians, over a period of three years his mathematical output was prodigious.

Early in 1917 Ramanujan fell seriously ill. (Though he subsequently recovered sufficiently to return to India in 1919, he was to die in April, 1920.) Hardy went to visit him in hospital, and for want of something to say, remarked that the taxi in which he had come had the number 1729 which seemed to be a rather "uninteresting" number. At once Ramanujan replied that, on the contrary, it was a very interesting number indeed, being the first one that is expressible as a sum of two cubes in two different ways.

The vast majority of us, even those of us who are professional mathematicians, can but marvel at a brain that can spot something like that. In fact, even if we are told the fact, checking it would prove quite a task for most people. Unless, of course, we have a micro sitting to hand. Then it becomes a pleasant exercise in computer programming. Find the two representations of 1729 as sums of two cubes and check that 1729 is indeed the first number with this property. (In this context, a "cube" is the cube of any whole number.)

Apparently Hardy went on to ask Ramanujan if he knew of a similar phenomenon for fourth powers, that is, did he know of a number N which is expressible as a sum of two numbers raised to the fourth power (i.e., multiplied by themselves three times) in two different ways. Ramanujan replied that he did not. (Remember, he was ill at the time!) In fact there are such numbers. Can you find any? This could involve a lot of computer time, so let me give you a little help. What you need to do is search for numbers A, B, C, D such that

$$A^4 + B^4 = C^4 + D^4.$$

You can find such numbers A, B, C, D all less than 160. A standard FOR-NEXT loop should now do the trick. I shall give a solution to this problem some time in the future.

While I am on the subject of Ramanujan, another example of his amazing abilities with numbers concerns the number π. This number is defined to be the ratio of the circumference of any circle to its diameter. It is known to be an irrational number, which means that in order to express its value accurately you need infinitely many decimal places. The infinite decimal expression for π begins: 3.141 592 653 589....

School textbooks usually take as a reasonable approximation to π the quantity 22/7, but this is only correct to two decimal places (22/7 = 3.142...). Ramanujan "found" the approximation $355/113 \times (1 - 0.0003/3533)$. This is correct to an astonishing 15 places of decimals, and is certainly adequate for most applications of π in calculations. It can be used to give an "accurate" value of π when multi-word arithmetic is being used in a computer calculation.

48

Square deals

April 11, 1985

A *multigrade* consists of two collections of numbers whose sum is the same and whose squares have the same sum. For example, the two groups 1, 6, 8 and 2, 4, 9 form a multigrade, because

$$1 + 6 + 8 = 2 + 4 + 9$$

and

$$1^2 + 6^2 + 8^2 = 2^2 + 4^2 + 9^2.$$

Starting with one multigrade, it is easy to create many others. You can add a fixed constant C to all the numbers and the result is still a multigrade. Or you can multiply all the numbers by a constant C.

Alternatively, you can start from scratch. Take any simple equality, such as

$$1 + 4 = 2 + 3.$$

Add any constant number to each term to give a new equality. For example, if we take the constant 4 we get

$$5 + 8 = 6 + 7.$$

Now switch ends in the second equation and add it to the first:

$$1 + 4 + 6 + 7 = 2 + 3 + 5 + 8.$$

If you check you will find that you now have a multigrade:

$$1^2 + 4^2 + 6^2 + 7^2 = 2^2 + 3^2 + 5^2 + 8^2.$$

Try it for yourself for any numbers you like.

What about going higher than the second power? If we now call multigrades like the above "second order" ones, are there any third order ones? The answer is yes. The equation

$$1^n + 6^n + 11^n + 16^n = 2^n + 4^n + 13^n + 15^n$$

is valid for n equal to any of 1, 2, and 3. Such higher order multigrades are also preserved by adding or multiplying by a constant, so again once you have one you can make many more. Or you can construct them from scratch. How? I'll leave you to figure that out.

The next thing to try is to make what might be called multi-multigrades, where you have three sets (or more) of numbers. For example, the equation

$$9^n + 25^n + 26^n = 10^n + 21^n + 29^n$$
$$= 11^n + 19^n + 30^n$$

for n equal to both 1 and 2.

How far can this kind of thing be continued? Well, that's up to you to find out. It is obviously a good problem for a micro attack. I'd be interested to find out how anyone gets on. I cannot claim to know much about all of this. Most of the above information was sent to me by Donald Cross of the University of Exeter, who has written several articles on the topic.

Another problem that looks ripe for a computer solution (though it can be done without, as I am sure many of you will tell me) is to arrange the digits 0 to 9 to make the following multiplication work out correctly:

$$** * \times ** = ** * **$$

where the very last asterisk has to be the digit 1.

This one is, of course, somewhat similar to the one about inserting plus and minus signs into the digits 1 to 9 in order to make a correct sum to 100, such as

$$123 - 45 - 67 + 89 = 100.$$

This is a problem I have mentioned before on this page. There are ten solutions besides the one quoted above. There are a variety of other problems of this nature.

Finally, a problem for the running fraternity. At a recent race in which there must have been somewhere between 100 and 1,000 runners, I noticed that the sum of the numbers greater than mine was equal to the sum of the numbers less than mine. The numbers had, by the way, been assigned consecutively, starting from 1. What was my number? [See Chapter 51.]

49

Square dance

April 25, 1985

On September 30, 1983, in front of a packed audience at the DESY research institute in Germany, the arithmetical wizard Wim Klein was presented with the four-digit number 5462. Within 44 seconds he had managed to express this number as a sum of four squares in no less than 10 different ways. One of Klein's factorizations was

$$5462 = 73^2 + 9^2 + 6^2 + 4^2.$$

How long will it take you to find the other 10, even using a micro? How long would it take the micro once the program is written and stored?

Expressing numbers as sums of squares is a game with a long history. It is an easy task to show that there are numbers which cannot be expressed as the sum of three squares. (Incidentally, you are allowed to use 0 in your expressions.) Are four squares always enough? That the answer to this question is "yes" was first proved by the great French mathematician Joseph Lagrange, who in 1766 filled the post vacated by Euler when he left for St. Petersburg, namely the chair in mathematics in Berlin.

(That Lagrange should take Euler's position was suggested by the mathematician D'Alembert, to whom Frederick the Great wrote: "I am indebted for (your) having replaced a half-blind mathematician with a mathematician with both eyes, which will especially please the anatomical members of my academy.")

Because of Lagrange's theorem that every number is expressible as a sum of four squares, it makes sense to write a computer program which will actually obtain such an expression. The first thing you might like to do with such a program is to tabulate, for each number, how many different expressions of it there are of this kind.

A more general question of the same kind is to ask how many cubes you need to express a given number? Or fourth powers? And so on. In a book written in 1770 (so before Lagrange's work), the English mathematician Edward Waring stated that any number can be expressed as a sum of nine cubes and of 19 fourth powers. Waring based these assertions on rather shaky numerical evidence, and did not have a proper proof.

The first real result of this kind other than Lagrange's theorem was due to Liouville who, in 1859, showed that every number is a sum of 53 fourth powers. More recent work has suggested that Waring was nevertheless correct in saying that 19 fourth powers always suffice. (It is known that 18 do not. I do not recall seeing anywhere an example of a number not expressible as a sum of 18 fourth powers, and do not know if it is within easy reach on a micro.)

Waring was also correct in his claim that nine cubes are always enough. In fact there are only two numbers which require nine cubes; eight are enough for all others. Both numbers are well within the range of a micro, so I'll leave it to you to find them (and their expressions as sums of nine cubes).

Yet another similar question: can an Nth power ever be the sum of fewer than N Nth powers? For example, can a cube ever be the sum of two cubes? This particular case indicates that Fermat's Last Theorem is floating around in the background, which might put you off a bit. But, although Euler conjectured that no Nth power can be a sum of fewer than N Nth powers, there is a fifth power which is the sum of four fifth powers. It was discovered in 1968 by Lander and Parkin. Can any-

one find their example? The numbers themselves are quite small, but since you have to raise them to the fifth power, large numbers are involved in the calculation.

Questions like this have occupied numerous mathematicians through the ages. Though the questions themselves may seem somewhat esoteric and without any application, research into such problems has often led to important discoveries. For example, the extremely important subject known as "ideal theory" arose out of work in number theory, and recent work on computer algorithms for factorization (important in cryptography) stems from similar problems.

50

The world would end before you could answer the questions

May 9, 1985

The most important single question in computation theory today is whether or not P and NP are the same. Hands up all those who don't even know what the letters P and NP stand for in this context?

The problem is purely a theoretical one, but should it turn out that P and NP are equal, the repercussions throughout computer science could be tremendous. In particular, most of the present day techniques of message encryption would be rendered highly vulnerable, for their security depends on an assumption that P and NP are quite different.

So what are P and NP? It's all to do with the speed at which computer programs can run. This depends upon a number of factors.

First there is the speed of the computer used to run the program. For the purposes of the P versus NP problem, this is not important: differences in running speeds can be ignored as "negligible." (You will see why one can cheerfully ignore a factor of a million or so—a factor that could probably only be achieved at immense cost to the computer manufacturer—in a moment.)

Then there is the skill of the programmer. This can be decisive, though it does not affect the problem I'm about to describe.

Finally there is the task the program is designed to carry out. This is what is at issue as far as P and NP are concerned.

In order to explain the problem, I'll take as an example one of the most famous computational problems around, the Travelling Salesman Problem, described in this column last December. [See Chapter 40.] A salesman has a list of cities he must visit. He has an atlas which tells him the distance between each pair of cities on this list. He wants to work out the route he must follow in order to minimize his total journey distance.

Can he write a computer program which will do this for him? The answer is, he probably cannot; very likely no one ever could. (At least, if P and NP are different they could not!) The critical factor is the time it would take for such a program to produce an answer.

Suppose we were to write a program to work out the salesman's route, using whatever computer we have access to. Given the program and the computer, the time taken to find an answer would, it would seem, depend on the number of cities to be visited. The more cities to handle, the longer it would take the program to find an answer. A crude measure of the efficiency of our program would be to express the time taken to solve the problem as a function of the number of cities involved.

In a general context, the relationship between the running time of a program and the size of the data is called the "time complexity function" for the program. If N is used to denote the size of the data and T the time complexity function, the quantity $T(N)$ tells us how long the program takes to find an answer given an amount N of data.

A problem is said to be of type P (for "polynomial time solvable") if there can be a program to solve it which has a time complexity function T which is bounded above by some polynomial functions (i.e., for some number k, $T(N)$ is not greater than N^k, for all values of N). It should be stressed here that this is not a question of which programs there are available today, but rather of any program which could ever be written. (Mathematicians always think big!) As far as we know, there is no program of the above sort which would solve the Travelling Salesman Problem, so this problem is probably not of type P. (Existing packages for this problem provide acceptable approximate solutions, not optimal ones.)

However, it is not hard to imagine a hypothetical "computer" which could solve this particular problem very easily. The difficulty in the problem is the sheer number of alternative routes available even for quite modest numbers of cities. (For N cities there are $N!$ routes.) Suppose we could somehow lay our hands on an infinite number of identical computers, and program each one of them to examine just one possible route for any proposed salesman's tour, and then compare the results (in one go) and thereby arrive at the shortest.

Since the problem of calculating the length of any one tour is easy (it's a simple addition sum), our "supercomputer" would solve the problem in no time at all. (You would need infinitely many computers to allow for arbitrarily large proposed tours!) Problems are said to be of type NP (for "non-deterministic polynomial time solvable") if they can be solved by running a polynomial time program simultaneously on an infinite number of identical computers and comparing all the answers in one go at the end. For obvious reasons, this concept is a purely theoretical one, but, as it turns out, a very important one.

Practically all of the present day problems to do with route allocation, scheduling, optimization, and the like, are of type NP. In order to be solved (exactly) on a computer they would have to be of type P. But does it follow from the fact that a problem is of type NP that it is in fact of type P? Probably not, but no one has been able to settle the question as to whether P and NP are actually the same one way or the other.

Instead, the main overall development to come out of all the work on this problem has been to up the ante to a daunting extent. It is now known that practically all of the "important" problems in NP (including the Travelling Salesman Problem) are what is known as "NP-complete."

This means that the problem cannot be solved in polynomial time *unless every other NP problem can be so solved.* Most workers regard the discovery that their problem is NP-complete to indicate that it is time to look for something else to work on. The only hope of solving such a problem would be to show that P and NP are identical. The consensus view is that they are not, despite the absence of any concrete proof of this fact so far.

Incidentally, in case you are wondering just what is the significance of using polynomials (expressions like N^k) to measure the efficiency of programs, consider the following figures. Suppose we have a computer capable of performing one million operations per second. A program with time complexity function $T(N) = N^2$ might require 0.001 seconds with data of size 10 and 0.0025 seconds for data of size 50. A program for which $T(N) = 2^N$ would require 0.001 seconds for the size 10 data but 35.7 years (!) for the size 50 data. With data of size 60 the figures are 0.0036 seconds and 366 centuries, respectively.

Most real NP-complete problems are "slower" than 2^N and involve more data than 60 units, and the time required to solve them exceeds the time span of the universe. This is why the polynomial/non-polynomial split is so crucial, and explains why the P and NP problem does not depend upon actual computer speeds.

51

Printouts and the negative computer

May 23, 1985

Part of this week's column is concerned with providing an update on problems in previous weeks.

On April 11 [Chapter 48], I asked what number I was assigned in a race where the sum of the numbers less than mine was the same as the sum of the numbers greater, the field being between 100 and 1,000, with numbers assigned consecutively starting from 1. My number was 204, with a total field of 288 runners, a result many readers found, and one that requires only elementary school algebra.

The piece on Ramanujan on March 28 [Chapter 47] generated a larger postbag than usual. I stated Ramanujan's observation that 1729 is the smallest number that can be expressed as a sum of two cubes in two different ways. These are

$$1729 = 12^3 + 1^3 = 9^3 + 10^3.$$

I suggested that you should use your micro to check that the number 1729 is the least such, and in fact many of you did this using a micro. But you don't need to use a computer to do this. As Denis Ward of Reigate pointed out, since there are only 12 numbers whose cubes are less than 1729, by examining all sums of pairs from this list which might conceivably produce an example smaller than 1729, it is a simple matter to verify that there are in fact none less than Ramanujan's number.

Mr. Ward went on in fine style to dispose of another problem I posed, this time on April 25 [Chapter 49]; namely to find a number which cannot be expressed as a sum of fewer than 19 fourth powers. The answer is 79, as you can easily verify for yourself. (A classic theorem of mathematics says that every number can be expressed as a sum of 19 fourth powers—at least if the number has fewer than 310 or more than 1409 digits!)

Also in answer to a problem posed in that column, the only two numbers which cannot be expressed as a sum of fewer than 9 cubes are 23 and 239; all other numbers can be written as the sum of 8 cubes.

Many readers used a micro to try to find four numbers A, B, C, D such that

$$A^4 + B^4 = C^4 + D^4.$$

I hinted that there were such numbers below 160, and this information can be exploited to speed up an otherwise horrendously time-consuming search, as several of you discovered. Mrs. R. Roby of Liverpool went one better. Without the aid of a micro, she finally managed to come up with the numbers 59, 133, 134, 158, which satisfy the stated requirements.

Going back even further, on March 14 [Chapter 46], I wrote about negadecimal numbers, using negative number bases, and I mentioned that I thought that negative base arithmetic had been considered for computer design, providing as it does the advantage of an arithmetic where there is no need for a negative sign.

Professor W. M. Turski, visiting Imperial College London from Poland, wrote to tell me that a computer was actually built which used "−2" base numbers both in the arithmetic and in the memory addressing. It was the UMC-1, a Polish-made computer of the late 1950s and early 1960s, of which several dozen were made and installed. Does anyone know of any other early computers using non-standard kinds of arithmetic?

52

How long is a coastline?

June 20, 1985

When this column began almost two years ago. I decided that when I had a spare moment I would write something about the fascinating new mathematical theory of "fractals." That spare moment never seemed to arrive; there was always something else to write about. But a rash of recent articles on the subject has persuaded me to make up for what must by now seem like a glaring omission.

First of all there was an article on this very page earlier this year describing a new computer game called *Rescue on Fractalus* brought out by Atari in conjunction with Lucasfilms, a game which depends upon fractals for its graphical display. Then there was the news that fractals were used to produce the graphics in the film *Return of the Jedi*, another Lucasfilms production. (In fact, film producer George Lucas has now what must be one of the strongest computer graphics groups in the world, utilizing the most powerful modern supercomputers to generate incredibly realistic computer displays.)

In March, the new journal *Micromath* appeared, including a highly readable account of the subject of fractals by Richard Noss. Then in its April 4 issue, *New Scientist* provided us with another excellent account of the subject by Michael Batty.

The modern theory of fractals began with a paper by French mathematician Benoit Mandelbrot in 1967, entitled "How Long Is The Coast of Britain?" Mandelbrot's point was that this question does not have a unique answer, or if it does it must be "infinity." Here's why.

Viewed from outer space, the coastline looks fairly smooth, and knowledge of the scaling factor would enable you to calculate the length.

But if you are in an airplane flying along the coastline at, say 20,000 feet, you would see many indentations that were not visible from out in space. Allowing for these promontories and coves would double or treble the value you get for the length of the coastline.

If you now follow the coastline on foot, you would notice still more detailed fluctuations in the coastline, and these would again increase the value you get.

Get down on your hands and knees and still more detailed fluctuations appear.

Use a microscope and there is more again.

Theoretically there is no end to this process. (In practice the answer you get would presumably depend upon some arbitrary definition as to where you stop looking for more detail.) The coastline is an example of a "fractal curve."

Using a micro it is not hard to produce some simple examples of fractals. (You could also use paper and pencil. The micro just does the job quicker and more accurately.) Start by drawing a straight line one unit long. Then re-

FIGURE 17.   Koch's Island. This snowflake-like island has an infinitely long shoreline.

place the middle third of the line by the three sides of a square drawn on this third. Then do the same with each of the five straight lines in this figure, and so on. Theoretically you continue indefinitely. In practice four or five iterations result in a picture where any further alterations are too small to notice. The diagram shows what happens for the first two repetitions.

Ideally, since a procedure like the one just described involves repeating the same operation over and over again, it is best handled using a language which allows you to call a procedure from within that procedure.

BASIC does not allow this, but Logo and Pascal do. BASIC programmers will have to get around this problem by writing several identical sub-routines, but since four or five repetitions of the generating process should be enough, this need not be too onerous.

A particularly attractive fractal is obtained by starting with a triangle and repeatedly replacing middle thirds by triangular pieces to produce the "snowflake curve" shown in the picture. Experimentation will result in a whole variety of attractive and highly symmetrical patters. Mathematically, fractals are interesting because they are figures whose dimensions are fractional. Whereas the usual lines and curves of elementary geometry are one-dimensional, planes and surfaces are two-dimensional, and so on (!) the "square fractal" described above has dimension about 1.5 and the snowflake fractal has the dimension about 1.26. (The exact values are $\log 5/\log 3$ and $\log 4/\log 3$.)

When you come to think about it, fractals are more common in the real world than are the "smooth" curves of classical mathematics. Look at your trusty ruler under a microscope if you don't believe me. In his excellent book *The Fractal Geometry of Nature* (Freeman, 1982), Mandelbrot gives scores of examples of fractals in nature. In fact, he gives so many instances that I found the book impossible to read in the normal way; it just overwhelms. I had to resort to occasional browsing, for which it is well suited.

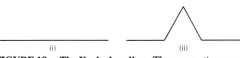

FIGURE 18.   The Koch shoreline. The generation process for the shoreline of the Koch snowflake consists of endlessly repeating this simple "middle-third" modification. Starting with a triangle, each such step, when repeated on every stretch of coastline, increases the total length of the coastline by a factor of 4/3.

FIGURE 19.   Koch's Island taking shape. The first three of the infinitely many stages in the production of Koch's snowflake island.

53

Bringing back beauty from the frontiers of chaos

July 4, 1985

Two weeks ago, while regular readers of this column were learning about the mathematical theory of fractals. I was in Oxford looking at some of the most impressive examples of what might be called fractal art that I have ever seen, in an exhibition organized by the Goethe Institute. This worldwide travelling exhibit of the work of H. O. Peitgen and P. H. Richter of the Center of Complex Dynamics at the University of Bremen in Germany is currently touring Britain.

Compared with the pictures on display at this exhibition, the kind of regular fractals which I described last time [Chapter 52] seem tame stuff indeed. The fractal-like structures involved in the exhibition involve chaotic behavior of one kind of another, a fact indicated by the exhibition's title *Schoenheit Im Chaos,* which the organizers have mysteriously translated as *Frontiers of Chaos,* an altogether much less satisfactory name. For beauty there is in great abundance. If you have never really believed that there can be beauty in mathematics then go to this exhibition and see for yourself. Modern techniques of color computer graphics bring a life and beauty to mathematical processes which will surprise seasoned mathematicians, let alone the man in the street.

If you go, take £4 so that you can buy the book which goes with the exhibition (and which has the same title, and is published by Mapart of the University of Bremen), a lavishly produced color glossy containing many more pictures than are in the version of the exhibition that I saw, and a book which struck me as grossly underpriced.* Besides all the amazing pictures, this book tells you about the mathematical theory behind the exhibition pictures, right down to instructions enabling you to produce examples of your own (though on a home micro you are not likely to achieve the quality of the pictures on display).

What is the mathematics involved? The buzz words are "complex dynamics," a relatively new area, currently receiving a lot of attention in the research world. The area abounds with rich mathematical structures, and computer graphics are used to visualize these structures, providing workers in the field with a valuable research tool and the rest of mankind with some fascinating pieces of art. To say any more (and I will not be able to say much here) I need a simple example.

If you take the simple mathematical formula x^2 and start to apply it to some number in an iterative fashion, three distinct types of behavior can result. If you start with the num-

---

* Now available from Springer-Verlag under the title *The Beauty of Fractals.* No longer as cheap, but still well worth buying.

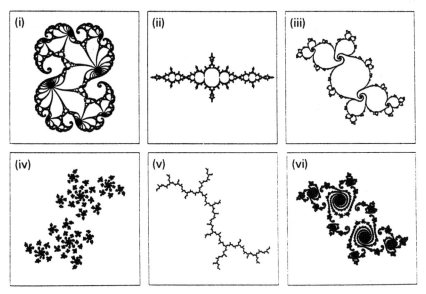

FIGURE 20.   Fractals. These particular fractals are known as Julia sets.

ber 0.1, when you apply the function to this
you get $0.01 (= 0.1^2)$, then when you apply the
same function to this number you get 0.0001,
then 0.00000001, and so on.

The sequence of values you get converges
towards zero. In fact, whenever you start with
a number whose absolute value (i.e., the value
ignoring any minus sign) is less than 1, your sequence of values will always converge towards
zero. Zero is an "attractor" in the region of
numbers of absolute value less than 1.

If, on the other hand, you start with the
value 1, then repeated squaring produces no
new numbers; you keep on getting 1. The
starting value of -1 produces 1 when you
square it the first time and thereafter repeated
squaring again keeps producing 1. The numbers 1 and -1 are "frontier" points. If you
start with any number of absolute value bigger than 1 the numbers you get tend towards
infinity (i.e., increase without bound); "infinity" is the attractor in this region.

Though very simple, the above example already contains the main ingredients for a rich
theory, one which is of fundamental impor-

tance in many areas of mathematics (analysis, approximation theory, and statistics to
name just three having obvious use in the real
world), and in engineering, biology, and psychology (and probably other subjects as well).

The basic idea is that you have some process involving a feedback mechanism and you
look at what happens when this process is applied to various starting points. The questions
of central interest are: Are there any attractors, and if so where are they? Over what
regions do the attractors act as attractors?
Within the region of influence of an attractor,
how fast is the convergence towards that attractor (e.g., how many iterations of the process are required to reduce the distance from
the attractor by a factor of 10 or 100)?

For the simple example of the process of
squaring a real number I gave above, these
questions are easily answered and the answers are of no great interest. But if you make
things even slightly more complicated the situation changes drastically. For instance, what
happens if you look instead at the function
$y = x^2 + c$, where c is some fixed constant, and

where instead of restricting yourself to real numbers x, y, c you allow complex numbers (numbers of the form $a + ib$ where i is the square root of -1)? Already you need computer graphics to provide a visualization of what is going on.

By looking at examples like the one above involving complex numbers, and varying the various parameters, you obtain the kinds of picture shown in the exhibition. Colors are used to provide contour maps showing aspects such as the "distance" (i.e., number of iterations required to reach it) of a point from an attractor. In this way an order begins to ap-

pear amidst the apparent chaos of the complexity of the situation. By varying parameters in a regular fashion, the computer can produce a "movie" of the mathematics as it takes place.

Ten such movies have been collected together into a video which runs continually during the exhibition. As a mathematician I would have liked a commentary with this, saying a little about the mathematics behind each sequence (the video just has an electronic music soundtrack), but even without it, the video is totally captivating.

54

A fractional approach to the pursuit of an ideal

July 18, 1985

To the mathematician, one of the most fundamental classifications of the ordinary, positive numbers is into the two classes of rational and irrational numbers, the former being those which can be specified by means of a finite or recurring decimal expansion, such as $3/8 = 0.375$ or $2/3 = 0.666...$ (recurring), the irrational numbers being those for which you need infinitely many decimal places, not falling in a repetitive pattern, such as $\pi = 3.14159...$ (ad infinitum). In everyday life, it is the rational numbers that play the major role. For most of present-day mathematics, the totality of all real numbers, both rational or irrational, is required, and since the turn of the century it has been known that the infinitude of irrational numbers is "greater" than that of the rational numbers.

But decimal expansions are not the only way of specifying numbers. One particularly interesting way (which, as it happens, has played an important role in recent years in connection with the security of secret codes) can be traced back to Rafael Bombelli, a mathematician of Renaissance Italy, who, in 1579 developed a theory of so-called "continued fractions." (I am being a bit provocative here. Three centuries earlier, Fibonacci considered a similar concept, and I dare say it could be argued that the whole thing goes back to the Hindus, but Bombelli was the one who really got the modern theory moving.)

The idea is a simple one, best illustrated by an example. The continued fraction representation of the number $3\frac{21}{26}$ is:

$$3 + \cfrac{1}{1 + \cfrac{1}{4 + \cfrac{1}{5}}}$$

The idea is to write the number as a nested sequence of fractions each with numerator 1. If you multiply the above expression up, you will see that it is equal to $3\frac{21}{26}$. Because it is so clumsy to write out a continued fraction in the way I have above, it is customary to capitalize on the fact that it is only the leading whole number and the successive denominators which are relevant, by writing, in the case of the example give, $[3; 1, 4, 5]$. This is what is called the "continued fraction expansion" of the number $3\frac{21}{26}$.

It is quite easy to write a program to calculate the continued fraction expansion of a given number X. Just set up a loop to calculate the three sequences A, B, C of numbers according to the rules:

$$A(O) = \text{INT}(X);$$

$$B(O) = X - A(O);$$

$$C(O) = 1/B(O);$$

$$A(N + 1) = \text{INT}(C(N));$$

129

$$B(N+1) = C(N) - A(N+1);$$
$$C(N+1) = 1/B(N+1).$$

(Check that $B(N+1)$ is not zero before this last step. If it is, you have reached the end of the expansion). Then the expansion of X is $[A(O); A(1), A(2), \ldots]$. Rational numbers have finite continued fraction expansions, irrational ones have infinite expansions.

And of course it is also easy to write a program which converts a finite continued fraction expansion into a fraction.

Some irrational numbers, which cannot be completely described using decimal expansions, have an easily remembered continued fraction expansion. For example, the number

$$e = 2.71828\ldots,$$

which forms the basis of natural logarithms, gives the recognizable pattern

$$e = [2; 1, 2, 1, 1, 4, 1, 1, 6, 1, 1, 8, \ldots]$$

where the even numbers appear in turn, separated by two 1s each time. Two more nice examples are

$$(e-1)/(e+1) = [0; 2, 6, 10, 14, 18, \ldots]$$

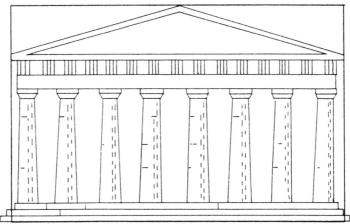

FIGURE 21.　The Golden Ratio. Greek architecture, as exemplified by the Parthenon in Athens, is said to incorporate the golden ratio.

and

$$(e^2 - 1)/(e^2 + 1) = [0; 1, 3, 5, 7, 9, \ldots].$$

Unfortunately, π does not yield such a nice pattern; it starts off

$$\pi = [3; 7, 15, 1, 292, \ldots].$$

The fact that there is perhaps more to continued fractions than just an odd way of representing numbers is indicated not only by examples like the ones above involving the number e, but by many other instances of "famous" numbers having remarkable continued fraction expansions. For instance, what number do you think is represented by the unending expansion

$$[1; 1, 1, 1, 1, \ldots]?$$

The answer is the well known "Golden Section" number $\frac{1}{2}(1+\sqrt{5})$, the so-called ratio of "ideal proportions," where a line is divided so that the ratio of the smaller to the larger is equal to the ratio of the larger to the whole. The story goes that the Greeks regarded this ratio as the ideal one for the design of an aesthetically pleasing building, though to be frank it has never struck me as being anything special at all to look at. To my mind, the Golden Ratio (about 1.618) is too oblong. A much more natural proportion is provided by a piece of standard A4 paper, which my measurement tells me has a ratio of approximately 1.41.

And in case you are wondering, I calculate the *Guardian* to have the proportion 1.5385. Perhaps some of you schoolteachers out there could persuade your charges to see just what kinds of ratios other everyday objects give rise to. Let me know what they come up with, and I'll come back to it at a future date.

Postscript.   See Chapter 57 for more on the topic of A4 paper.

55

The pure delight of the mathematical pay-off

August 8, 1985

A little while ago I overheard part of a conversation between two academics, both in mathematically-based subjects. "Of course," said one dismissively, "it might be nice to do research in pure mathematics, but we should concentrate most of our efforts onto the more applicable mathematical subjects."

Whether or not the said individual really believed what he was saying I do not know. In the current climate in many British universities, with everyone clamoring for a slice of the rapidly disappearing cake, he may have been simply trying to secure his own corner. But I have a suspicion that there was more to it than that. In which case something somewhere has gone drastically wrong. Given the rather impenetrable nature of most present-day mathematics, one might expect such a view from a total layman, and from Sir Keith Joseph* I would expect nothing else, but from someone whose work depended upon the pure mathematics developed yesterday it is little short of incredible.

The truth is that, in a university context, exactly the opposite is the case. Given that it takes something like 10 years to train a professional research mathematician, during which time the attrition rate is quite high, the sensible course would be to encourage (or at least allow) the few who make it through this apprenticeship to get on with it. For it seems that, whereas many mathematicians are able to apply existing techniques to solve real-life problems, relatively few manage to succeed in pure research. The few that do are almost exclusively in universities.

"Ah yes," I hear you saying, "but there are only limited resources and we simply cannot afford to support the luxury of pure research." I would answer that, particularly in the context of Great Britain, we cannot afford to neglect the necessity of pure research. The problem is, and always has been, how can you illustrate the usefulness of a subject that hardly anyone understands? But occasionally the message does seem to get through. Where do you think the following quotation first appeared?

"As a group, mathematicians are among the most fascinating people we know. They have a quirky turn of mind that makes them interested in mathematics and at the same time makes them interested in unusual things that you'd otherwise never hear of . . . Besides being delightful, mathematics is also crucial to progress in science and technology. It is one of the most productive ways in which public money can be spent."

---

* The British government's Education Minister at the time.

So where did this appear? A mathematical journal? A submission to the University Grants Committee from a beleaguered mathematics department? The answer is, it is an extract from an editorial in the *Los Angeles Times* on January 15 of this year. If what the *LA Times* editorial writer says strikes you as pretty hard to swallow, then the fact is, I'm afraid, that you have no idea what mathematics is and what its practitioners are. In which case, the *LA Times* has something more to say to you, this time on November 28 last:

"For some reason, not altogether clear, many otherwise educated people delight in rolling their eyes and proclaiming proudly, 'I don't know anything about mathematics.' Would they be so proud of their ignorance in literature, for example, or art?"

Faced with the task of providing examples of concrete applications of *their* art, mathematicians scarcely know where to begin. Practically every piece of mathematics which has ever been produced has found a use sooner or later. What more abstract and obscure branch of mathematics could there be than mathematical logic, complete with its pages of squiggles and symbols? And yet it was this very subject which led to the invention and development of the stored program computer that we have all grown to love (?) of late. The theoretical possibility of such a device was first established by the logician Alan Turing in the 1930s. It was taken to an implementable form by Turing and another logician, the American John von Neumann.

Want another example? In 1940, the great English mathematician G. H. Hardy wrote:

"I have never done anything 'useful.' No discovery of mine has made or is likely to make directly or indirectly, for good or ill, the least difference to the amenity of the world."

Within forty years, Hardy's work would prove to be central to the design and operation of message encryption systems, as has been mentioned several times in this column. But in Hardy's words we begin to glimpse part of the problem with the image of mathematics as "not worth support."

Of late this situation has shown signs of change. Faced with assaults from several quarters, mathematicians are slowly coming to realize that in an age of PR, image counts. A few years ago the London Mathematical Society instigated an annual series of Popular Lectures, where leading mathematicians could be trotted out for public scrutiny. The Joint Mathematical Council of the United Kingdom set up an Images Subcommittee to see what could be done to improve the tarnished image of the subject. One or two totally unscrupulous individuals have even used the current interest in microcomputers as an excuse to spread the word through newspapers and magazines. And one brave BBC producer even had the audacity to display these "most fascinating people" on prime time television last year.* The mathematical empire is beginning to strike back.

---

* Keith Devlin collaborated with the BBC's Jon Palfreman to produce *A Mathematical Mystery Tour.*

56

Big guns go west

August 23, 1985

The most useful device for short-term data storage in existence today is undoubtedly the pending tray. By making use of the one on my editor's desk, I was able to ensure that while readers of this column were not deprived of their regular fix of mathematical diversions over the past few weeks, I myself was 5,760 miles away, or so the Pan Am pilot informed me. I was taking a glimpse into the future. As you might expect, the future is alive and well and living in California.

The object of my trip was to attend an international meeting designed to try to direct the way forward in computer science beyond the Fifth Generation and on into the Sixth, Seventh, and possibly the Eighth. (Who knows? The intellectual air in California is pretty heady). The venue for the enterprise was Stanford University, from whose picturesque campus came many of the early pioneers of the microchip revolution, leading to the creation of the famous Silicon Valley, which stretches from Stanford all the way down to San Jose at the Southern end of the San Francisco peninsula. (A glance through the local telephone directory reads like a Who's Who of computer firms. Accordingly a glance through the local newspaper reveals that it costs about $1,500 a month to rent a small house in the area.)

The original idea for the meeting—which was attended by over 250 people from around

the world—came from a research institute called CSLI (The Center for the Study of Language and Information) which was set up at Stanford two years ago with the aim of developing the theoretical tools that will be required to take us beyond the present highly crude techniques in computing and information technology.

It turned out to be the most amazing and exciting meeting I have ever attended. The speakers (and their audiences) consisted of mathematicians, philosophers, logicians, computer scientists, computer engineers, linguists, and who knows what else.

One thing that became clear from the start was how recent developments in computer technology had turned these disciplines onto their heads. For years, mathematicians, philosophers and logicians, though each dealing with essentially the same fundamental concepts, have gone their own highly disparate ways, with barely a glance at what each of the others was doing. Linguistics was regarded as a totally unrelated subject altogether, and as for computer science, at one time that was something that mathematicians (myself included) sent their less successful students away to do.

Now all this has changed, and anyone who wants to be involved in the future developments in computing (in its wider sense) will need to know about all of these things. In his stimulating and entertaining introductory address, the Director of CSLI, the philosopher John Perry, joked that over the past couple of years he and his colleagues in philosophy had discovered that their brand of "obscure intellectualism" had suddenly become "highly fundable when carried out in front of a computer terminal." It was a good joke which went down well. It was also very true.

So just why is everyone looking towards everyone else for help in solving their problems? A simple example might give you some idea. Computing is, as everyone knows nowadays, about handling data (information), numerical or otherwise. Computers work by moving data around in certain ways (determined by a program), and conversely computers are used to transfer data from one person/place to another. But when you look at it, the whole question about how information is transferred from one person to another (or one computer to another) is extremely complex and not well understood as yet.

Suppose I say to you "It is raining." What information has passed from me to you? Certainly not the information that it is raining. For one thing, I may be speaking to you on the telephone, and so it might be raining where I am but not where you are; in which case which location am I talking about? Or I may be lying or mistaken. There is a whole range of possibilities, and to decide between them you need to somehow combine my assertion with other information, in fact with much more information than is carried by me saying the original sentence. The only information that you gain from my saying "It is raining" is that I said it. Even if you know that I always tell the truth, all that you can conclude is that I believe it to be raining. You might like to figure out what extra information is required before you really can conclude that it is raining. It's not easy. When you have done it, ask yourself how a computer could be programmed to

make the same deduction. (Don't spend too long on this. At the present it is beyond our capabilities. But we're working on it!)

If a seemingly simple example like the one above turns out to be so complicated, what about the problem of trying to design systems to help with medical diagnoses or run nuclear defense systems? At the moment we are light years away from a solution; we rely on guesswork and intuition, items which are all too often wrong. So, increasingly, computer scientists are looking to philosophers, linguists, and cognitive psychologists for help in designing and building their systems.

In the other direction, linguists are using computers to help analyze language and its use as a carrier of information, and philosophers and psychologists are looking at the way computer systems make decisions to see if they can gain any insight into how the human mind works. In short, a whole range of quite different fields are rapidly coming together.

At the Stanford meeting there were talks from (among many others) the computer scientist John McCarthy on Artificial Intelligence (the subject which owes its name to him), by the linguist Barbara Partee on mathematical theories of human language, by the Swedish philosopher Per Martin-Lof on the question as to what constitutes a mathematical proof, by the computer scientist Larry Wos on an "intelligent system" developed at the Argonne National Laboratory, and by the computer engineer Stanley Rosenchein about the work being done at Silicon Valley's own SRI on the design and construction of robots.

It was a heady mixture, even without the added attraction of the coincidental presence on the Stanford campus of a convention of football cheerleaders. Many new and potentially powerful contacts have been made between the various disciplines (discounting the cheerleaders), and we are unlikely to have seen the last meeting such as this.

In fact, the three weeks at Stanford were so stimulating and encouraging that it seems a pity to finish on a sour note. But as I write this column I am preparing to return to the UK where the atmosphere is entirely different. When CSLI was set up two years ago it was given a grant of $25 million to see it through the first nine years. Already there are over 100 people attached to the institute, which now has links with various hardware and software companies in the area. Yet if the Stanford meeting had been held just a couple months later I for one would have been unable to attend. The latest round of government cuts in university research expenditure caused the curtailment of the research fund which paid my expenses to California. All in all it seems an odd way to run a country. Perhaps one day we shall be able to lease from the Americans a computer system which, when I type in the sentence "Sir Keith Joseph* runs the education system" is intelligent enough to recognize that there is a letter "i" missing in the fourth word.

---

* The British government's Education Minister at the time.

57

How the beauty of mathematics brought a sense of proportion to origami

September 5, 1985

Hands up all those who know the mathematical principle behind the A-range of stationery? Readers of this column on July 18 [Chapter 54] will be well aware that I, for one, did not. In their droves they were quick to point out that what I took to be a somewhat arbitrary choice of paper sizes was in fact based on sound mathematical principles.

For new readers the story begins like this. An oft-repeated statement to be found in popular books on mathematics is that the rectangle which the human eye finds the most aesthetically pleasing has its sides in the proportion known as the Golden Ratio, whose value is approximately 1.618, and of which more in a moment. In my July 18 article I said that this particular ratio had never struck me as anything special to look at, and my skepticism that the Ancient Greeks really used this proportion as a basis for their architecture was confirmed by the Warwick-based historian of mathematics David Fowler. My claim was that a piece of standard A4 paper provides a much more pleasing shape. I measured the ratio in this case to be 1.41.

It turned out that my aesthetic sense corresponded to a mathematical fact. The sides of a piece of A4 paper are in fact in the proportion of $\sqrt{2}$: 1, i.e., a ratio of 1.414.... In fact the whole range of A-sized paper is based on this ratio. The largest size available, A0, is designed to have an area of 1 sq. meter. If you

fold that in half you get two pieces of A1-sized paper. Fold a piece of A1 in half to get A2, and so on down to A5, which thus has an area of 1/32 sq. meters. All foldings are along the shorter direction.

What makes the A-sizes special is the requirement that the ratio of the sides for each member in the range should be the same (i.e., they should all be the same shape). The only proportion which will give this is $\sqrt{2}$: 1, as a little elementary algebra will indicate.

So, A0 paper has dimensions 1189mm by 841mm (to give an area of 1 sq. meter) and A4 is 297mm by 210mm. And there, in a nutshell, is a classic example of my oft-repeated claim that there is beauty in mathematics and vice versa. My experience also throws an interesting new light on a classic quotation from Omar Khayyam (in his treatise *Discussion of Difficulties in Euclid*): "Finding out about ratios is a secret of deep logic."

One curious aspect of the A-system is that it is theoretically impossible to give the exact lengths of the two sides of any member of the series. No matter what scale of measurement you use, however fine, you cannot make a piece of paper M units by N units so that the ratio of M to N is $\sqrt{2}$. (Mathematicians express this fact by saying that the number $\sqrt{2}$ is irrational.) Of course, this has no bearing on real life, where mathematically perfect accuracy is never important or achievable.

Getting back to the Golden Ratio, this appears at the beginning of Book 6 of Euclid's *Elements* (ca 350 B.C.). It is the ratio you get when you divide a line into two pieces so that the ratio of the whole line to the longer piece equals the ratio of the longer piece to the shorter. If the ratio concerned is $x : 1$, then what is required is a solution to the equation $(x + 1)/x = x/1$, that is $x^2 = x + 1$. School algebra gives the answer GR $= \frac{1}{2}(1 + \sqrt{5})$, which is about 1.618.

Putting the algebra into words, GR is the number which has the property that if you take away 1 and then reciprocate the result, you end up with the same number again.

The Golden Ratio crops up in other parts of mathematics. One well-known example is in connection with Fibonacci sequence. This is the sequence of numbers you get starting with 1 and forming the next number at each stage by adding together the two previous ones (except at step 2 where you only have one previous number). So the sequence begins: $1, 1, 2, 3, 5, 8, 13, 21, \ldots$. If $F(n)$ denotes the nth number of this sequence, then as n gets bigger the ratio $F(n + 1)/F(n)$ of successive terms of the Fibonacci sequence gets closer and closer to GR.

There is also an explicit formula which allows you to calculate $F(n)$ from n, and this also involves GR. Unfortunately it is a bit too messy looking to write out here.

To finish with, those of you who still use your micros (plus those who prefer more traditional methods) might like to have a look at the following problem. What do you notice about the following table of numbers?

1	9	2
3	8	4
5	7	6

First of all, each of the digits 1 through 9 appears once in the table. Second, the number on the second line is twice that on the first. Finally, the last number is the sum of the first two. How many other sets of three three-digit numbers can you find which have all three properties just mentioned? [See Chapter 61.]

58

How to take an electronic line for a walk

September 19, 1985

It was Bertrand Russell who said, in his 1918 book *Mysticism and Logic,* that: "Mathematics, rightly viewed, possesses not only truth, but supreme beauty—a beauty cold and austere, like that of sculpture." Anyone who succeeds in surviving school "mathematics," and goes on to find out what "real" mathematics is all about will know at once what Russell meant. For the vast majority who never manage to get far beyond the "Let x represent the unknown quantity" stage, the advent of the electronic computer has provided an alternative way of getting to see some of this beauty that is claimed to lie beneath the surface.

Just as television travelogues are never as good as being there for yourself, so computer graphics displays of mathematical phenomena never manage to capture all of the beauty in mathematics. (Russell was not just thinking of a beauty which could be visualized.) But they go a long way, and indeed they can provide the professional mathematician with insights that would not otherwise be available.

All of which is an excuse to show you another example of the computer graphics displays produced by the Complex Dynamics Research Group at the University of Bremen in the excellent book *Frontiers of Chaos,* which accompanies the travelling display of some of their work currently on show around this country. [Chapter 53].

The fundamental ideas behind the creation of such pictures is the repeated application (i.e., iteration) of some particular mathematical procedure, possibly coupled with some randomization. A particularly simple example which is easily implemented on a home micro is the following.

Fix some value of the constant A between 1 and 4. Pick some number x between 0 and 1. Now keep replacing x by the new value $A * x * (1 - x)$, plotting the value of x you get each time ($*$ means multiply by). So you keep feeding the old value of x in to get the new value, which overwrites the old value and is then itself fed back in to the process. By plotting the values of x as you go you get a visual record of what goes on. You will have to experiment with the scale on your display to get a picture of reasonable size.

If you try different initial values of x you will find that the end result is much the same, whatever the fixed value of A you chose in the first place. Much more interesting is what happens when you fix one value for x to start at and look at different values of A. To see what happens, you should plot values of A along a vertical axis from 1 and 4 and the values of x you get (starting at the same initial value for each choice of A) along the horizontal axis on a scale which gives you a good display. Let A increase in steps of, say, 0.1 in the first instance, and iterate the old x/new x procedure, say, 100 times for each A. (But experiment until you get a result which you like.)

Roughly speaking, what you will find is that, for all values of A up to 3, the plotted points rapidly congregate around one value. For A between 3 and around 3.57 you get successive (and increasingly rapid) splitting into more and more values around which the x-sequence jumps back and forth. (To improve the picture try just plotting the values of x after the first twenty or so—again experiment.)

Above $A = 3.57$ there is chaotic behavior where the x values jump all over the place. But as A increases still further order reappears:

at around 3.82 the x values start jumping between just three values. Then a bit further on chaos reigns once more.

Having discovered where the really interesting behavior occurs, do the whole thing again but concentrate on the range of values of A which you want to look at and increase A through this range in much smaller steps than before, using a different vertical scale so that you get a "magnification" of the area in question. Try plotting more values of x each time. Try missing out the first 100 values. Try whatever seems like a good idea. The beauty of the micro is that all of this experimentation is eas-

FIGURE 22. A fractal world.

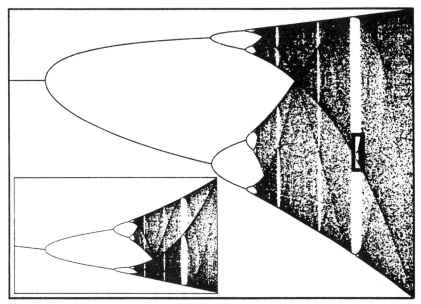

FIGURE 23.   The Verhulst process. Bifurcation in action.

ily performed, and if something does not work you have lost nothing except the odd micro-penny-worth of electricity.

When you have grown tired of this exercise, try a different procedure for generating new x from old. Don't write to me and say that the various values I suggested above are not the best. They are not intended to be. They are just to get you started. The real enjoyment in this kind of thing is discovering things for yourself.

Incidentally, the pictures of the kind illustrated (which are in color in the originals) are produced by iterating procedures using complex numbers. Since you need two dimensions to plot complex numbers, the effect of changing the parameter (the A of my example) is illustrated by color and shading, as in a contour map.

59

Factor factors

October 3, 1985

Every positive whole number is either a prime number or else is a product of prime numbers. This has been known since the work of Euclid around 350 B.C. So, for example, 3 is prime, 11 is prime, 30 is the product of the primes 2, 3, 5. How to find the primes which go together to make up a given number is thought to be one of the most difficult problems in computational mathematics. For fairly small numbers there is no problem. A straightforward search through all the primes less than the square root of the number in question will work. But for a number with, say, 50 digits this procedure could take a billion years, even using the most powerful computer available.

Though the factorization problem has been worked on by one or two mathematicians over the years (including the famous French mathematician Pierre de Fermat), it was only the recent discovery that the difficulty of factoring numbers could be used to design highly efficient encryption schemes that turned the factorization problem into a major topic of research. With so much effort being put into the problem, something was bound to give and it just has. The Dutch mathematician Hendrik Lenstra recently announced that he had found a relatively simple yet effective method of factoring to add to the batch of other techniques which have been developed since the advent of the computer age.

Lenstra's method will not effect the cryptographic community (it appears), since they depend upon the difficulty of factoring a large number (say of 160 digits) which is known to be the product of two prime numbers of about the same size (around 80 digits each).

The new method works best only with numbers which have prime factors of different sizes. (For numbers with equal-sized factors the method does work, but no better than the other available techniques, which broadly speaking will factor numbers of up to around 70 digits as a matter of routine and bigger numbers if you are lucky.)

Though simple and efficient, Lenstra's method uses some sophisticated mathematical concepts involving "elliptic curves" in order to find the factors. But it is to some extent comparable with an earlier method due to the English mathematician Pollard, which is easy enough to try out on a home micro. Instead of using elliptic curves, Pollard's method utilizes simple quadratic functions such as $x^2 + 1$. In order to try to factorize a number N, you start with some number $X(0)$ less than N and define a sequence of numbers by the rule

$$X(I + 1) = X(I)^2 + 1 \pmod{N}.$$

As you go, you keep working out the highest common factor of N and each of the numbers $X(I) - X(J)$ for all values of J less than I (I alters as well). This can be done using a highly efficient technique known as the Euclidean Algorithm, details of which can be found in any elementary book on number theory or computational techniques involving numbers. (Briefly, to find the highest common factor of A and B define

$$R(1) = A \bmod B,$$

$$R(2) = B \bmod R(1),$$

and then

$$R(I + 2) = R(I) \bmod R(I + 1)$$

until you get a value of R which is zero. The previous value of R is then the highest common factor of A and B.)

Most of the time the highest common factor of N and $X(I) - X(J)$ will be 1. But fairly soon you should get a value greater than 1. When you do, this highest common factor will be a factor of N and you have done what you set out to.

The theory behind this method does not show that you are guaranteed an answer, only that it is extremely unlikely that you will fail. But I have tried this method myself, and it does work extremely well on numbers of 30 digits (which require special programming) on a small computer. The method has been used to factor numbers of around 80 digits on large computers.

60

The perfect picture

November 14, 1985

What do you do with the painting that refuses to hang square? In a research paper in the journal *American Mathematical Monthly* last May, F. J. Bloore and H. R. Morton of Liverpool University tackled this whole thorny issue. In spite of the apparent simplicity of the problem, the solution turns out to involve some pretty sophisticated mathematics, involving such mathematical notions as vector equations, evolutes, and cusp catastrophes. Some idea of the nature of the paper can be gleaned from the sentence which the authors use to commence their investigation. "It is fruitful to consider the locus of the hook rela-

tive to the equilibrium position of the picture as the cord length is varied," they claim, and so indeed it turns out to be the case, though for most citizens the sentence is more likely to bring back long forgotten memories of afternoons spent wrestling with a school geometry textbook.

In a nutshell, what the paper demonstrates is this. Imagine drawing the diagonals joining the two pairs of opposite corners of the picture. The angle between them measured about the three o'clock position is called the "critical angle." If the angle made by the cord holding the picture at the hook is greater than

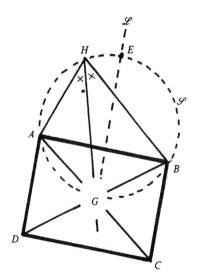 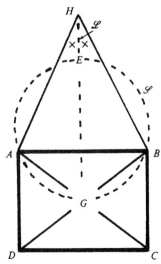

FIGURE 24. How to hang a picture; and how not to—the mathematical story.

the critical angle, the picture will be in an unsuitable position when hung and is likely to end up tilting. If the angle at the hook is less than the critical angle, the picture will hang straight.

So the secret is to make the cord long enough. This could cause quite a problem. As the authors point out, if you wanted to hang the Bayeux tapestry (assuming it were fastened to a board) you would need a cord one kilometer long. (Being 1/2 m by 17 m long, the tapestry has a very small critical angle.)

In a further section to the paper, the authors show that, if the cord is not attached to the picture centrally, you will have to contend with troublesome things called strophoids, which I don't want to go into here. The upshot is that in this case it is impossible to hang the picture properly. So now you know.

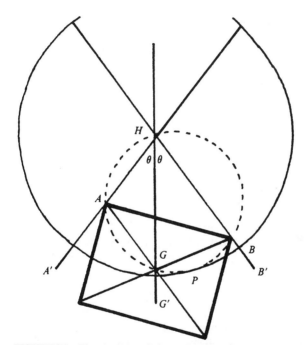

FIGURE 25.   The geometry of picture hanging.

61

Quite a performance

December 5, 1985

It stands just 45 inches tall and is shaped in the form of a 300 degree arc of a circle 53 inches in diameter. It has four background processors each with their own high speed local memory, and a common memory capable of storing 256 million 64 bit words. Its international state changes every 4.1 billionths of a second. Its circular arc is divided into 14 vertical columns each of which contains 24 modules weighing 2 lbs apiece and using 500 watts of power. Because of the high density with which the silicon circuits are packed into these modules, the entire machine requires a sophisticated cooling system in which the fluorocarbon cooling liquid is in direct contact with the electrical components. 2,000 lbs of the monolith's 5,500 lb weight consists of this coolant.

The Cray-2 computer, is the newest and fastest of a line of supercomputers produced by computer engineer Seymour Cray. According to the advertising literature, the new machine operates 4 times faster than its predecessor, the Cray XMP (itself an enhanced Cray-1). But just how fast does this make it?

The experts will tell you at once that all talk of "computing speed" is pretty meaningless. It all depends on the calculation you want to perform. Some machines perform well at one type of calculation, others shine in quite different areas. For example, a colleague of mine recently performed some tests on the Cyber 205, a supercomputer built by Control Data. After some experimentation he managed to write a routine which ran on the computer at a speed within half a percent of the manufacturer's stated top speed.

To do this he had to instruct the machine to perform a highly repetitious and essentially useless arithmetic calculation, unlikely to arise often in practice. This does not mean that there is anything wrong with the Cyber 205.

But those same experts who state quite correctly that numerical comparisons of computer speeds are not relevant, devise and run tests on the various machines and publish the results in a league table of top computers. One such comparison table, published by the Argonne National Laboratory in the United States, is in front of me now.

The table is based on tests using standard Fortran software for handling linear equations, a fairly common task for high performance computers. All results are given in terms of "Cray equivalents," i.e., how many computers of the type considered does it take to equal the performance of the Cray-1 as far as the task in hand is concerned.

The Cray XMP mentioned above scores best; it only takes 0.36 of one to equal a Cray-1. (The table was drawn up before the Cray-2 was available.) The Cyber 205 comes next at 0.48. The Cray -1S scores 0.54. Then come the Japanese with the Fujitsu VP-200 at 0.64. IBM only comes in at 23rd place with their 370/195 at 4.9, but then they never did get really involved in the supercomputer game, preparing instead to dominate the huge middle ground of mainframe computing.

The Vax 11/780 which I and hundreds of other university-based researchers use weighs in at 76 to the Cray, closely followed at long last by a British machine, the ICL 2988 scoring 85 to gain 65th place in the table. Down at the bottom of the league we find the IBM PC, of which you will need 21,875 to equal a Cray. With a Cray system costing somewhere in the region of $20 million, the question is largely academic for most users.

Finally, here is the solution to a problem I set some time ago [Chapter 57] which does not require a Cray-2 for its solution. I gave the table of numbers

$$1\ 9\ 2$$
$$3\ 8\ 4$$
$$5\ 7\ 6$$

which has the properties that each digit from 1 to 9 is used once and once only, the second number is twice the first, and the third is the sum of the first two. I asked for other examples of tables with these three properties. There are three such:

$$219, 438, 657; \quad 273, 546, 819; \quad 327, 654, 981.$$

- *Time*'s Woman of the Year:

  Corazon Aquino

- Challenger Explodes

- Chernobyl

- US bombs Tripoli

- *Crocodile Dundee*

- Fergie and Andy

- Racketeer John Gotti

  goes to trial

- Mets take World Series

- *Top Gun*

- Bullwinkle moose loves

  Jessica cow

- "Baby Doc" Duvalier

  leaves Haiti

- Ivan "The Terrible" Boesky

- Clint for Mayor

- Iran-Contra scandal gets

  special prosecutor

- *Aliens*

- *The Fly*

- *Hannah and her Sisters*

- Black Thursday: Dow drops

  86.61 pts

1986

62

In pursuit of prime suspects
January 9, 1986

If you were to write down 1031 successive 1s, what would you get? The answer is a prime number. And not just any old prime number at that, but a record breaking prime number.

Mathematicians have been interested in prime numbers since the days of the ancient Greeks. Euclid proved that there are infinitely many prime numbers, that is numbers which cannot be divided exactly by any smaller number other than 1. This in itself may appear counterintuitive at first sight. For while there are plenty of prime numbers at first—2, 3, 5, 7, 11, 13, and so on—the bigger the numbers you look at, the less common the primes become. Below 10 there are four primes, between 10 and 20 there are only two, and likewise between 20 and 30. Higher up they are even sparser, so that it is not at all obvious that they don't eventually simply peter out altogether. But Euclid's result demonstrated that this does not in fact occur.

One of the most basic results of mathematics, so important that it is given the impressive sounding name of the Fundamental Theorem of Arithmetic, is that every positive whole number can be decomposed in a unique way into a product of prime numbers (except where the number is prime already, of course). This means that for the mathematician interested in whole numbers the prime numbers are like the chemist's elements or the physicist's fundamental particles. An immense amount of effort is put into the study of prime numbers.

In spite of the apparent simplicity of the task, checking to see if a given number is a prime can be a major challenge unless the number is fairly small. The most obvious method is to try to divide the number by all smaller numbers. If this proves to be impossible, then the number is a prime. But for anything but very small numbers this procedure will obviously take a great deal of time and effort. It can be speeded up by observing that as it stands the process involves a lot of repetition. It is enough to trial divide by all smaller prime numbers, and in fact only those prime numbers less than the square root of the number being tested. But even this will not enable a test to be made on numbers of 30 or more digits. For instance, for a number with 50 digits it would take the most powerful computer available 100 billion years to test for primality using trial division. Using sophisticated alternative mathematical techniques it would take the same machine about 15 seconds.

It is in the development of these highly efficient primality tests that a lot of the present interest in hunting for record prime numbers lies. The so-called *repunit numbers* provide one hunting ground. These are numbers which, as their name suggests, consist of the digit 1 repeated a certain number of times. The repunit consisting of n 1s is denoted by

R_n. Thus R_1 is 1, R_2 is 11, R_3 is 111, and so on. There are obviously infinitely many repunits. But how many of them are prime? The answer is not many. To date only five are known.

The first repunit prime is R_2, i.e., 11. At this point the problem becomes too great for a home micro. The second repunit prime is R_{19}, followed by R_{23}. Until recently these were the only ones known. Then, about seven years ago, Hugh Williams, of the University of Manitoba, showed that R_{317} was prime. He also showed that R_{1031} was probably prime.

How did he do that? Well, there are some methods available for testing primality which, though extremely fast, don't provide a 100% reliable answer. These methods are called "pseudo-prime tests." A number that fails the test will definitely be non-prime. A number that passes the test is highly likely to be prime, in the sense that there are relatively few non-primes that pass the test.

So, since 1979 it had been suspected that R_{1031} was prime. It was a long struggle to get a proper answer. Only this year did Williams, aided by Harvey Dubner of Computer Systems Inc., obtain the proof that this number really is prime. It took 20 days of computing and a lot of mathematical sophistication to do it.

It may also be the largest one ever discovered. For Dubner has shown that if there is another repunit prime (and no one knows if this is the case) it has to be greater than $R_{10,000}$, which probably puts it out of range of even supercomputers. Unless, of course, someone discovers another way of testing for the primality of repunit numbers. And that is not as unlikely as you might think. For numbers having special forms it is sometimes possible to devise highly specialized primality tests which capitalize on that form.

Postscript.   See Chapter 85 for more on repunit primes.

63

A monk whose mathematical genius was almost infallible

January 16, 1986

Last week [Chapter 62] I discussed the recent discovery of Hugh Williams and Harvey Dubner that the number consisting of precisely 1031 digits 1 is prime. I mentioned that, in order to test the primality of numbers with many more than 30 digits, use has to be made of sophisticated mathematical methods, trial division by all primes less than the square root of the number being much too slow even on a fast computer. But even highly efficient tests begin to struggle with numbers of a thousand or more digits. The ARCL Test, probably the fastest general purpose test available, which can handle a 50-digit number in around 15 seconds on a Cray-1 as against 100 billion years using trial division, will require a week's computing time for a 1000-digit number.

For numbers too large to handle using general purpose tests, special techniques are required. These make use of whatever facts are known about the number. For instance, for the 'repunit' number mentioned at the start of the article, the primality test run by Williams and Dubner makes explicit use of the fact that the number consists of a string of 1s.

Numbers which are one less than a power of 2 can be tested for primality by means of an extremely fast test known as the Lucas–Lehmer Test. For many years now the largest known prime numbers have all been of this form. A few weeks ago the record fell again.

The new holders of the somewhat dubious title of discoverer of the world's biggest prime number are Chevron Geosciences in Houston, Texas.

Their Cray XMP computer took three hours to show that the number $2^{216091} - 1$ is prime. Written out in standard decimal format this number has 65,050 digits, beginning with 7460 and ending with 8447.

Though the computer age has provided some measure of fame for numbers of the form $2^n - 1$, the credit for the discovery that such numbers are quite interesting goes back to a seventeenth century French monk called Marin Mersenne, and these numbers are nowadays known as *Mersenne numbers*. In his book *Cognitata Physica Mathematica*, written in 1644, Mersenne made some pretty startling claims about Mersenne numbers which are prime. He claimed that $2^n - 1$ is prime for the values of n equal to 2, 3, 5, 7, 13, 19, 31, 67, 127, 257, and fails to be prime for all other values of n less than 257. In fact he was not quite correct. In 1947, using desk calculators, it was discovered that taking n to be 67 and 257 does not give a Mersenne prime, while n equal to 61, 89, and 107 does. But given that Mersenne did not have any kind of mechanical aids, his prediction is astonishingly good, which suggests that he may well have made use of some fast primality test which enabled him to perform the necessary calculations, but which contained a logical flaw leading to his erroneous conclusions.

The Lucas–Lehmer Test is sufficiently simple to enable the home micro user to try it out, though in order to deal with reasonably large examples it is necessary to write special arithmetic routines to handle numbers which are stored as arrays rather than in single words. To check if a number $2^N - 1$ is prime, calculate the sequence of numbers $U(0)$, $U(1)$, up to $U(N - 2)$ using the rules:

$$U(0) = 4;$$

$$U(K+1) = (U(K)*U(K)-2) \bmod (2^N-1).$$

If, at the end, $U(N - 2)$ is not zero then $2^N - 1$ is not a prime. (Though you have no idea what any of the factors of this number are, of course.)

For instance, to use the test on the number $2^5 - 1$ (which, being 31, we know is a prime before we begin), the numbers $U(0)$ up to $U(3)$ you get are 4, 14 mod 31 = 14, 194 mod 31 = 8, and 62 mod 31 = 0. This last value of zero is what tells you that the number $2^5 - 1$ is prime.

The world record prime number mentioned above is the 30th Mersenne prime which has been discovered to date. It is not known whether there is an infinite number of Mersenne primes, nor indeed if there are any Mersenne primes between the new one and the last one found, $2^{132049} - 1$, which was discovered in 1983 using a computer like the one used this time. The reason for this gap in our knowledge is that there are so many possible candidates to look at in searching for record primes that statistical arguments are used to try to predict where the next one may be, after which the search is concentrated around that region. As always with such imprecise arguments, it is always possible that there is a rogue example which does not obey the "laws" of statistics.

Postscript. See also Chapters 1, 2, 4, 8, 105, and 130.

64

As easy as pi
February 27, 1986

"How I wish I could enumerate pi easily which men known skillful calculate forever..." That opening sentence was not written by me, but by an American called Jay Jung, as reported in the October, 1985, issue of *Scientific American*. In fact, the complete sentence has 111 words. What makes it special is that if you list the numbers of letters in each of the words in the sentence you obtain the decimal expansion of π, that tantalizing infinite sequence which begins:

3.14159 26535 89793 23846....

Another π mnemonic quoted in the article is that of Peter Brigham: "How I wish I could enumerate pi easily, since all these (censored) mnemonics prevent recalling any of pi's sequence more simply." I must confess Brigham's sentence appeals to me, and I expect I will always be able to recite the first 20 places as a result. The nice thing about it is that it actually makes sense in relation to the task it is dealing with, namely remembering π. That makes it far better than the classic physicist's mnemonic: "How I want a drink, alcoholic of course, after the heavy chapters involving quantum mechanics."

There are similar mnemonics for the mathematical constant e, the base for natural logarithms, which starts off:

2.71828 18284 59045 23536....

Of course, provided you are not too bothered about accuracy there are plenty of other ways to remember π. For instance, π is approximately equal to the number of seconds in a year divided by 10 million. (In other words there are approximately π times ten million seconds in a year.) And π is quite well approximated by the square root of 10.

For yet another way to get an approximate value for π, try entering the number 2143 on your calculator, divide by 22, and then press the square root button twice.

Using powerful supercomputers, π has been calculated accurately to at least 10 million places in recent years, but until the electronic age mankind had made little progress on this task for many centuries. Making direct use of the definition of π as the ratio of any circle to its diameter (modern techniques are different), Archimedes (ca. 200 B.C.) made the first calculation of π, concluding that it lies between the two fractions 22/7 and 223/71, the former being the classic "schoolboy value." In fact 22/7 differs from π in the third decimal place. A much better value is 333/106, correct to the fourth place, and 355/113 which is out by only 3 ten millionths or so. This latter value was known to the early Chinese mathematician Tsu Chung-Chih around A.D. 500.

In the sixteenth century, a German called Ludolph van Ceulen calculated π to 35 places and had the result carved into his tombstone. This feat seemed to spark off a small wave of

π calculators, culminating in a certain William Shanks spending most of his life calculating the wretched number to 707 places sometime in the nineteenth century, fortunately passing away long before desk calculators came into use and and an error was found in the 527th (and subsequent) place, leaving Shanks only slightly ahead of the 500 places obtained by Richter in 1854.

Besides providing grist for the mill when it comes to testing out new computers (and calculation of π to several million places can take 24 hours on a modern supercomputer), the unending expansion of π gives people with good memories something to try their minds on: memorizing the expansion. The current record holder is an Indian, Rajan Srinivasen Mahadevan, who recited 31,811 places in 1981. An English youth, Creighton Carvello, managed just over 20,000 places in 1980.

I cannot remember if these records have been broken since then.

65

A prime target

March 27, 1986

Two American mathematicians have recently discovered a new and potentially very fast method for testing large numbers to see if they are prime, a problem which became significant outside mathematics when sophisticated encryption systems were developed whose security depended upon the availability of an efficient prime testing method.

A prime number is a whole number that cannot be divided (exactly, leaving no remainder) by any smaller number other than 1. Thus 2, 3, 5, 7 are primes while 4, 6, 8, 9 are not. Though for small values it appears that primes are very common, as the size of the numbers increase they become less frequent. There are only 168 primes below 1,000, for example.

Prime numbers are important to the mathematician because, as was known to the ancient Greeks, every whole number is made up of primes multiplied together, which makes primes the mathematical analogue of the chemist's elements or the physicist's atomic particles. The cryptographic applications of primes mentioned above make them of strategic importance to large security concerns as well. In the past decade, an immense amount of work has gone into developing efficient methods for dealing with primes on a computer.

Testing a given number to see if it is prime is easy enough provided the number is small: three or four digits if you are doing it by hand, or about ten if you have a computer. All you need to do is try to divide your number by all primes less than its square root. If none of these divides your number, you know it is prime. If one does, it cannot be prime.

The difficulty with this simple approach is that for larger numbers it becomes impractical. Even if you had access to the fastest computer in the world (with the best programmer as well, if you like), a 20-digit number could take two hours to check if it was prime, while a 50-digit number could take 10 billion years. Since the cryptographic applications of primes require primes of around 150 digits or more, obviously other methods are used.

There are a number of alternative ways to test for primality of a number. In the early 1970s, Michael Rabin of the Hebrew University in Jerusalem and, independently, Robert Solovay of the University of California at Berkeley (then visiting IBM) and Volker Strassen of the University of Zurich discovered two similar probabilistic ways of testing for primality. Though very fast and practicable, they suffered from the drawback of there being a small but real chance of coming up with the wrong answer.

If either of their tests said your number was not prime then you could be sure of the result. But if the test decided that your number was prime it was always just possible that it was not. At the cost of a small amount of extra computation time, the chance of an error could be made as small as you pleased, but would never be zero.

Another method, potentially the fastest of all, was designed by Gary Miller of the University of Southern California. Though its answer can be relied upon when it comes up with "not prime," a question mark hangs over all "prime" responses. The problem here was not a probabilistic but a mathematical one. The test depended on a famous conjecture about complex numbers called the Extended Riemann Hypothesis. Since this hypothesis has not yet been proved, Miller's test thus stands on an uncertain foundation. (A rumor that the regular Riemann Hypothesis was solved last year [Chapters 39 and 41] turned out to be false.)

A totally secure method was finally developed by Leonard Adleman and Robert Rumely in 1980, and subsequently improved by a number of other mathematicians so that current computer implementations routinely handle numbers of 200 digits or so in a matter of seconds. But for significantly larger numbers even this test begins to lose efficiency in an ever more rapid fashion.

Ideally what is required is a test which is always right and whose efficiency does not wane as the size of the numbers goes up. (Of course, the bigger the numbers, the longer you expect the computation to take. What is at issue here is speed relative to the size of the number.) Miller's test would be such a test provided the assumptions it is based on were proved. The Adleman–Rumely test is definitely not of this type.

The new test, developed recently by Shafi Goldwasser and Joseph Kilian of the Massachusetts Institute of Technology, involves a degree of uncertainty, but not in the sense that its decisions might be wrong. Whatever the test tells you can be relied on. The uncertainty lies in the fact that although the test is in general fast and efficient, there is a small chance that it will run slowly (for certain primes). To just how many primes, if any, this would apply is not at present known, though the available evidence suggests that such primes would be rare.

The test uses some very new mathematical ideas involving elliptic curves, which number theorists are finding an extremely powerful tool. Last year Hendrik Lenstra at Berkeley used elliptic curves to develop a new method for factoring large numbers.

66

Maths can be good for you

April 24, 1986

Does mathematics really have very much to do with computing? Apparently a lot of people do not think so. Frequently, university admissions officers tell of applicants having been advised to concentrate on the "useful" subjects like computing and engineering, and to avoid the "less relevant" mathematics. To the mathematician, such a view seems incredible, but since it appears to be a growing one there must be some reason for it.

The root of the problem is nicely summed up by what Paul Newman says just before being blasted away in the final scene of the film *Cool Hand Luke* (the one about the chain gang): "What we have here is a failure to communicate." The message about mathematics is simply not getting through. Why? Partly it has to do with the nature of mathematics. It is, without a doubt a particularly impenetrable subject, even to most other scientists let alone the man in the street. And yet until recently it was widely accepted as being useful. So what has changed?

The answer is, at least in part, that the baby to which mathematics gave birth a couple of decades ago is beginning to grow up. Not only is computer science now a subject in its own right, but its very nature has made it a big-budget discipline. With universities struggling to stay alive, money has become a dominant factor in deciding which subjects ought to be encouraged and which left to get by as best they can. In this climate, computing departments find themselves in a very powerful position, while the relatively small-spending mathematics departments do not. The baby has eclipsed the parent.

Yet the need for computER science is no longer as great as it was. What is required urgently is computING science. It is not so much the machines themselves that require study but how to use them. This point is emphasized rather nicely in an article by the re-knowned computer scientist Edsgar Dijkstra in the magazine *Mathematical Intelligencer* last month. As Dijkstra points out, when you go into a hospital for an operation, you hope to be dealt with by someone who has learned how to use a knife, not someone who has spent five years studying the design and manufacture of the implement itself.

The computer scientists have done a good job. The machines are now there and they work pretty well. Practically all the major problems that remain involve the proper use of those machines. Both at the research level, and even more at the teaching level, nearly all the emphasis ought now to be on the use of computers. This is the point Dijkstra makes in his article. And, as he says, what you need in order to make efficient use of computers is—wait for it—mathematics.

Dijkstra is by no means alone in this view. In the United States, a committee set up to design a model syllabus for university computing courses came to the same conclusion. What was needed was mathematics and lots of it. (The findings of the committee were published in two reports in the journal *Communications of the Association for Computing Machinery*, 1984, pp. 998–1001, and 1985, pp. 815–818). The amount of mathematics that is recommended in this report is far greater than is currently included in most computing courses offered both in this country and the USA. It is also largely of a type not traditionally given a great deal of emphasis in university mathematics departments.

In the computer age, the heavy dependence on the calculus-based mathematics that still forms the bulk of most degree courses has begun to appear decidedly anachronistic. What most students want (need) today is what is known as discrete mathematics—or rather more of it than they are currently getting. So

it is not just the computing departments that have to change, the mathematicians have to change too.

And there you have a problem. In times of financial crisis—which is the state most universities are in practically all the time at present—there is a tendency to retreat behind whatever defenses you can find. Jealous of their new-found independence from mathematicians, the computer people are suspicious (and possibly even afraid) of their former colleagues with three thousand years of history behind them, while for their part the mathematicians often resent the way that the new upstart seems to take practically all the available cake. Just when the greatest gains could be made by cooperation, circumstances seem to be forcing the two sides to become rivals. It is to be hoped that this state of affairs proves to be a temporary one.

67

Selling under false colors
May 8, 1986

"I'm sorry, sir, there's been a computer error." In the early days of computer technology that excuse would have satisfied most people. But now we are all so knowledgeable about what computers can and cannot do that few officials would dare to make such a remark. Computers do not make errors: people do. The computer simply does precisely what we tell it to do.

If the man in the street, or the local Gas Board office, realizes that there is no such thing as a computer error, you would expect that the very people who developed the computer—the mathematicians—would be the first to acknowledge results produced using their brainchild. But this is not the way it has turned out.

In 1976, two American mathematicians at the University of Illinois used a computer to solve a long-standing open problem of mathematics. In June of that year, Kenneth Appel and Wolfgang Haken announced that they had proved the Four Color Theorem, a goal that mathematicians had sought since 1852.

There was a catch. Their proof depended upon extensive use of electronic computers—some 1200 hours of computation in fact. So great was the amount of work performed by the computer that there was no possibility of human mathematicians checking all the details, as is the case with practically every other new result in the most precise of all the sciences. This heavy dependence on computers initially led many mathematicians to refuse to acknowledge the claimed solution as such.

But, in mathematics as in other aspects of life, there is no standing in the way of progress, and the voices of opposition slowly began to fade. They did not, however, die out, and over the last three or four years there has been a number of rumors to the effect that the proof is not in fact correct—that there has been a computer error.

In response to the renewed doubts concerning their proof, Appel and Haken have written a spirited defense of their work in the magazine *Mathematical Intelligencer,* Volume 8 (1986), pp. 10–20.

The Four Color Problem has its origin in a letter written in October 1852 by the newly qualified mathematician Francis Guthrie to his brother Frederick, still a student (of Physics) at Francis' old college, University College London. Francis had noticed that it is possible to color in the counties of a map of England using only four colors, where the coloring is subject to the obviously desirable requirement that no two counties sharing a length of common boundary should be colored using the same color. Was this the case for all possible maps, he asked, or are there maps for which five colors are necessary?

Frederick passed the problem to his mathematics professor, the famous British mathematician Augustus de Morgan. De Morgan had no difficulty in proving that there are maps which require at least four colors, but was unable to answer Guthrie's question. Nor were any of his students, to whom de Morgan communicated the problem.

In 1879, just one year after the mathematician Arthur Cayley had publicized the problem by raising it at a meeting of the London Mathematical Society, one of the society's members, a London-based barrister called Alfred Bray Kempe, published a paper in which he claimed to prove the conjecture. But eleven years later, one Percy John Heawood found a mistake in the argument. Kempe's proof was adequate to demonstrate that five colors always suffice, but did no more than that. However, Kempe had certainly approached the problem in what was to turn out to be the correct manner.

He began by assuming that there was a map which required at least five colors to color it properly. It follows that there will be such a map having the fewest number of countries (amongst all such maps). Although there is no way of knowing anything about this map, Kempe tried to show that it would have to contain (somewhere among its possibly millions of countries) certain configurations of countries.

By examining these individual configurations one at a time, he hoped to show that in each case it would be possible to reduce the number of countries in the configuration without reducing the number of colors required to color the map. This would mean that there would be a map requiring five colors having fewer countries than the one started with. Since that had been taken to be one with the fewest number of countries, that would be a contradiction, and the conclusion would be that the assumption that a map existed requiring five colors was false: four colors must always be enough.

In his proof, Kempe used just five configurations, but the argument about reducing the number of countries was erroneous for one of them. The 1976 proof by Appel and Haken used the same approach but involved some 1,482 configurations. The detailed analysis of such a large number of maps could only be carried out using a computer. Indeed, they had used a computer in order to find many of these configurations.

The published proof consisted of some 50 pages of text and diagrams, a further 85 pages giving 2,500 more diagrams, 400 microfiche pages giving details of various parts of the proof, plus the results of some 1,200 hours of computing which had to be taken more or less at face value. Small wonder that the proof was originally greeted with some skepticism. That a proof of such length and complexity will contain errors is accepted by all, including the authors. The question is, are there any serious errors which render the proof worthless?

In the ten years since the proof first appeared no one has been able to find one. A West German student by the name of Ulrich Schmidt did find one error in 1981, but Appel and Haken were easily able to correct it. No others have been found, aside from small errors of the order of misprints. The onus would now seem to be on the doubters to prove their case. In their recent paper Appel and Haken have reaffirmed their confidence in the result, and the majority of mathematicians (myself included) would be content to accept that. But there is no doubt that with the advent of the computer-assisted proof, the nature of mathematics has changed. Now we have to believe, without always being able to check for ourselves.

The same applies to the Star Wars project, of course.

Postscript. The Four Color Problem is also discussed in Chapter 18.

68

A Farey story

May 22, 1986

Once upon a time there was a Farey. This particular Farey was a mathematician. And the legacy he left behind can provide the home micro user with a bit of harmless fun, as well as something very deep for the professional mathematician to ponder over. It is all to do with fractions.

If you can remember back to your schooldays, you may recall that a "proper" fraction is one of the form h/k where h and k are both positive whole numbers, h is smaller than k, and h and k have no common factor. So, 1/2, 3/4, 100/101 are all proper fractions, but 4/3 and 6/8 are not. In 1802, a mathematician called Haros looked at certain sequences of proper fractions and was able to prove some curious facts that he had observed them to have, but his work attracted little attention. Then, in 1816, a chap called Farey stated one of Haros' results (without proof) in an article that he published. Seeing this article, the great French mathematician Cauchy supplied the missing proof to the result quoted by Farey, and gave the name "Farey sequence" to the the type of fraction sequence that was involved. For every positive whole number N, there is a corresponding Farey sequence, denoted by $F(N)$. It consists of all those proper fractions whose denominator is at most N, together with the additional "fraction" (which is not really a fraction at all) 1/1, all written out in increasing order.

So, for example, $F(1)$ consists of the single number 1/1, while $F(2)$ has two numbers, 1/2 and 1/1. Then $F(3)$ is the sequence

$$\frac{1}{3}, \frac{1}{2}, \frac{2}{3}, \frac{1}{1},$$

$F(4)$ is

$$\frac{1}{4}, \frac{1}{3}, \frac{1}{2}, \frac{2}{3}, \frac{3}{4}, \frac{1}{1},$$

and $F(5)$ is

$$\frac{1}{5}, \frac{1}{4}, \frac{1}{3}, \frac{2}{5}, \frac{1}{2}, \frac{3}{5}, \frac{2}{3}, \frac{3}{4}, \frac{4}{5}, \frac{1}{1}.$$

It is not at all hard to program a computer to work out Farey sequences. (But remember, it is important that the numbers in the sequence appear in the right order, namely getting bigger as the sequence progresses.) When you have done this you can check (either by hand or using the computer) that all Farey sequences have the following two curious properties.

First, if you take any three successive fractions in a Farey sequence, say a/d, b/e, c/f, then the middle one is equal to the fraction you get by adding the numerators and denominators of the two outside ones and then cancelling any common factors, i.e.:

$$\frac{b}{e} = \frac{a+c}{d+f}.$$

For example, the 10th, 11th, and 12th members of $F(7)$ are 4/7, 3/5, and 2/3. And when you work it out, you get

$$\frac{4+2}{7+3} = \frac{6}{10} = \frac{3}{5}$$

The other fact is this. If a/c and b/d are any two successive fractions in a Farey sequence, then $bc - ad = 1$. Taking $F(7)$ as an example again, the 6th and 7th members are 1/3 and 2/5, and when you work it out you find that

$$(2 \times 3) - (1 \times 5) = 6 - 5 = 1.$$

Check both of these facts for as many Farey sequences as you like. The result is always the same. In fact, if you are up to it, you might even like to try to prove these two facts using algebra.

So far, all of this looks like harmless fun rather than deep mathematics. But as is so often the case, just beneath the surface lurk some decidedly nasty currents. Farey sequences are intimately related to what for the professional is the single most important unsolved problem of mathematics: The Riemann Problem, first posed by Bernhard Riemann in 1859.

Though generally stated in terms of complex numbers, the Riemann Problem affects those most basic of mathematical objects, the prime numbers, and if the problem were to be solved tomorrow we would at once know a lot more about the primes than we do at the moment. But, as was discovered by Franel and Landau in 1924, the Riemann Problem has an entirely equivalent formulation as a question about Farey sequences. I'll tell you what it is. It is a little bit technical, but if you work out a few examples for yourself you should be able to see what is going on.

If you take any Farey sequence, the fractions in it all represent real numbers between 0 and 1. If you draw a line to denote the real interval from 0 and 1 and mark the Farey fractions (for some Farey sequence) on it, it will be apparent that the points marked are not spaced equally along the line. For example, $F(4)$ consists of the numbers 1/4, 1/3, 1/2, 2/3, 3/4, 1/1, while an equal spread of six points between 0 and 1 would give the numbers 1/6, 1/3, 1/2, 2/3, 5/6, 1/1. (As with the Farey sequences themselves, you exclude 0 but count 1/1.) If you work out the total amount by which the Farey fractions "miss" the equally spaced points in this case you find that it is

$$\frac{1}{12} + 0 + 0 + 0 + \frac{1}{12} + 0 = \frac{1}{6}.$$

This figure of 1/6 is denoted by $D(4)$ (for "difference"). In general, $D(N)$ is the total amount by which the members in the Farey sequence $F(N)$ miss the equally-spaced points between 0 and 1.

If you were to draw a graph of the behavior of $D(N)$ as N gets larger, you would find that it begins to looks like a graph of the square root of N. But is this just an "illusion" that would fail to hold for extremely large values of N, beyond the range of the most powerful computer, or is it always the case?

When formulated properly, this question is equivalent to the Riemann Problem. Which shows that when you play about with Farey sequences on your home micro, you are just a stone's throw away from the biggest unsolved problem in mathematics. [See Chapters 39 and 41 for a discussion of the Riemann Hypothesis.]

69

Blooming numbers
June 5, 1986

Spring, they say, is when tulips come from Amsterdam. This year the Dutch city has been producing blooms of a different kind—numbers, very big numbers, the kind that can be produced only with the aid of the most powerful computers available.

The garden which has supplied them is the Center for Mathematics and Computer Science, a research institute of the Stichting Mathematisch Centrum. Founded in 1946, this establishment is a nonprofit institution sponsored by the Dutch government. Its aims are the promotion of mathematics, computer science, and their applications. Results coming out of the Amsterdam center in recent years have made many appearances in this column. Two more may now be added to the list.

The first is a new factorization. Factoring is one of the most basic problems of mathematics. It has been known since at least 350 B.C. that every positive whole number is either a prime number or else is a product of two or more prime numbers, and is such a product in only one way—except for changing the order in which the primes are multiplied together.

Over the last 20 years mathematicians have developed efficient techniques for testing a given whole number to see if it is prime or not. Using the obvious method of systematically searching for a possible exact divisor will work only for very small numbers. For, say, a

50-digit number such a naive approach could take a thousand billion years computing time on the world's most powerful computer.

Currently available "clever" tests would require at most 15 seconds on the same type of machine. Such tests do not, however, tell you what the prime factors are of a number that turns out to be non-prime (and hence a product of primes). The question is not without significance outside mathematics, since a currently used method of encrypting electronically transmitted information depends upon the difficulty of finding the actual prime factors of a number which is known to be a product of two 60-digit (or so) primes. (If a method could be discovered to find these factors, the encryption method would no longer be secure—finding the factors will allow an eavesdropper to decode the message.)

The best general-purpose factorization methods that have been developed can handle numbers of around 60 digits, in a few hours of computation on the fastest machines available, though larger numbers have been factored—usually by exploiting some additional information about the number.

Factorization experts have drawn up a list of non-prime numbers which have been found to present especial difficulties as far as trying to find their prime factors is concerned. One of these is the 72-digit number which consists of thirty-six 9s followed by thirty-five 0s followed by a 1.

Using a new variant of a now standard method known as quadratic sieving, due to Peter Montgomery, a team at the Amsterdam center were able to factor this number on May 13 of this year. It turns out to be a product of a 34-digit number and a 39-digit number (both of these numbers being prime). Quoting such monster numbers fully here would, of course, leave you wondering whether there was a misprint, so you will have to make do with the knowledge that the computation took something over four hours on a Control Data Cyber-205 computer, one of a small number of high-performance mainframe computers that are so powerful they are known as supercomputers.

The best comparable previous result was the factorization in 1984 of the number consisting of 71 successive 1s. This was done on another supercomputer, the Cray XMP-24 at the Los Alamos Laboratories in the USA, by a team from Sandia Labs (USA). This machine is faster than the Cyber-205 for such calculations, and the fact that it took nine and a half hours to complete the job is because the method used (also a variant of quadratic sieving) was less efficient than the new Montgomery approach.

These results in no way imply that modern computers are capable of factoring any 72-digit number. Factorization is a funny game (and I use the word "game" advisedly), where computing power, mathematical skill, experience, shrewd guesswork, and no small amount of luck all play a part. Some numbers larger than the two quoted have been factored using quite small computers. For instance, Robert Silverman ran eight SUN-3/75 minicomputers in parallel for a total of 1260 hours to factor a certain 81-digit number, and factored a 75-digit number using a method involving 235 hours on a VAX 8650 and 40 hours on a VAX 780.

The other result to come out of the Amsterdam center this spring concerns the way the prime numbers are distributed among the rest of the whole numbers, and involves the biggest number ever to arise in classical mathematics. As you might therefore expect, discussion of this other Amsterdam bloom will have to be postponed until next time.

70

Power games

June 19, 1986

Think of the number consisting of 1 followed by 34 zeros. (That is, the number ten thousand million million million million million.) Now try to imagine the number that you get when you take a 1 and follow it by that many zeros. (You cannot, of course, visualize it written out—the universe is not big enough.)

Now make a further leap of the imagination by thinking of the number consisting of 1 followed by that last gigantic number of zeros. In mathematical language, what you have got is the number 10 raised to the power (ten raised to the power (ten raised to the power 34)).

That number, give or take the odd million, actually occurred in a piece of genuine, old-fashioned mathematics. And by so doing became at once the holder of the title "the biggest number ever to occur in mathematics." (The rules of this particular game are that the number has to have a genuine significance. You cannot simply make up a large number and then invent some use for it.)

The number (a more "accurate" value of which is given by using the EXP function available on most micros or pocket calculators, namely EXP(EXP(EXP(EXP(7.705))))), but don't try to work it out!) cropped up in some 1955 work of the mathematician S. Skewes concerning the manner in which the prime numbers are distributed among the other whole numbers.

At first, there seem to be a lot of primes. But as you go up through all the whole numbers, the primes become more rare. Home micro users can easily write a small program which, for each whole number N in turn, counts the number of primes less than N. Below 1,000 there are 168, below 10,000 there are 1,229, and below 100,000 there are 9,592. The frequency of the occurrence of primes can be measured by working out the "relative density" up to each stage.

For example, below 1,000 the density of the primes is 168/1,000 that is 0.168. Below 10,000 the density is 0.1229, and below 100,000 it is 0.09592. So the density is dropping. The question is, is this decrease in the density an orderly one or a random one?

It is by no means a trivial question, for although the primes do become less frequent, the exact place where they crop up seems to follow no regular pattern at all. There are long runs of numbers where there are no primes, then suddenly there will be a whole rush of them, then maybe just one prime on its own, then a pair of two successive odd numbers that are both prime, and so on.

Only on the more global scale of looking at the overall, "statistical" density has some order been discerned. For the density decrease *is* an ordered one. The density of the primes below any number N is very nearly equal to $1/\log N$, where logarithms are taken to the base e (so-called "natural logarithms," e being a mathematical constant which is approximately equal to 2.718). This result was first proved in 1896 by Hadamard and de la Vallée Poussin, and is known as the Prime Number Theorem.

What the theorem tells you is that the number of primes less than N, which I shall denote by $P(N)$, is approximately equal to the number $N/\log N$. But there are better approximations. One such was discovered by the child prodigy Karl Frederick Gauss in 1791 (he was 14 at the time).

Known as $Li(N)$, you get Gauss's number (for any given whole number N) by integrating the function $1/\log x$ from 2 to N. For the value $N = 1,000$ you get $N/\log N = 145$ and $Li(N) = 178$, while $P(N)$ is (as mentioned earlier) 168. For $N = 10,000$ you get $N/\log N = 1,068$, $Li(N) = 1,246$ and $P(N) = 1,229$, while for $N = 100,000$ the figures are 8,686 for $N/\log N$ and 9,630 for $Li(N)$, with $P(N) = 9,592$.

The above figures indicate how much better is the Gauss $Li(N)$ formula. They also suggest that $Li(N)$ approximates $P(N)$ a little on the large side. But is this always the case? Or does there come a value of N for which $Li(N)$ is *smaller* than $P(N)$?

In 1914, the English mathematician Littlewood proved that numbers $Li(N)$ must oscillate above and below $P(N)$ infinitely many times. But his highly abstract proof gave no indication where the first N is for which $Li(N)$ is less than $P(N)$. Then, in 1955, Skewes showed that such a switch must occur before the huge number quoted at the start of this article. Since that time mathematicians have attempted to reduce this bound.

In 1966, Sherman Lehman showed that below the number 165 followed by 1,163 zeros there are 1-followed-by-500-zeros successive values of N for which $Li(N)$ is less than $P(N)$, thereby reducing the Skewes bound quite considerably. Then, earlier this year, Herman J. J. te Riele of the Amsterdam Center for Mathematics and Computer Science (an institute which was the subject of my last column, on June 5 [Chapter 69]) showed that below the number 669 followed by 368 zeros there are 1 followed by 180 zeros successive values of N for which the same thing happens.

This brings the Skewes bound down to something that can be written out on a single page. The result required several hours of computation on a Cyber-205 supercomputer —a device not, of course, available to Skewes (or to Lehman for that matter).

71

A playful approach to the bomb

July 3, 1986

I once shared an office with one of the men who helped to build the first hydrogen bomb. The year was 1979, and I was giving a course of lectures at the University of Colorado in Boulder. During my stay I used the office of the mathematician Stanislaw Ulam. Having already retired from his position as chairman of that department, Ulam was spending most of his time at the Los Alamos Research Laboratories, several hundred miles away in New Mexico, where he had retained an attachment ever since arriving there to work on the atomic bomb project during the Second World War. And so his office was temporarily free.

So what insights did I gain into the great man while I was there? Unfortunately, none whatsoever. Apart from the standard university issue furniture, the only item in the room when I arrived was a single mathematics text book (First Year Calculus) on the front flap of which he had scribbled his name. Which left me in the same position as anyone else who knew of Ulam by reputation alone. I had to rely on his books and other writings to learn of his thoughts and his work. But his story is one that speaks volumes for those in society who think they can predict what will be "useful" for a student to learn, and who automatically devalue anything that looks remotely like fun and games.

For Ulam was, for his entire career, a man who never tired of playing games—in his case mathematical games. For him, any mathematical problem was a game or puzzle to be attacked just like a chess problem. Some of his games have so far had little or no practical applications. Others, like the work he did on Teller's H-bomb project, could hardly have had more significance.

Because of the time when he grew up and developed his mathematical abilities, it is not surprising that a great deal of Ulam's work was concerned with computers, and many of his original thoughts on the problems associated with large scale computation have turned out to have been extremely pertinent as the machinery has developed to make his ideas relevant. Though Ulam died two years ago, a recently published collection of many of these ideas enables anyone who so wishes to share in the thoughts of this great man.

Science, Computers and People: From the Tree of Mathematics, by Stanislaw Ulam, published by Birkhäuser Boston in 1986, contains both mathematical observations and historical accounts of parts of his life and that of others around him. For anyone whose interests in computing go beyond the latest video games, it makes fascinating reading. For as one of those few individuals who was actively involved with the very first development in computer technology, Ulam is able to supply a perspective that few others ever could.

Most of the new book concerns Ulam's work after he arrived in America from his native Poland, and in particular his work on computational problems. In a short article such as this it is difficult to even begin to give the flavor of what is in the collection—you will have to read it for yourself. But here is just one example of a typical Ulam "game" that can be played on any home micro—which will, of course, be much more powerful than the machines Ulam himself had available for most of his career.

Generate a sequence of whole numbers by the following simple rule. Start with the numbers 1 and 2. Then put in the sequence just those numbers which cannot be obtained as a sum of two previous numbers in the sequence in more than one way. Thus 3 follows 2 in the sequence, and the next number is 4, but then comes 6, because $5 = 1 + 4 = 2 + 3$. Likewise $7 (= 1+6 = 3+4)$ is left out, but 8 is included.

If you work it out, the first few numbers of the sequence are 1, 2, 3, 4, 6, 8, 11, 13, 16, 18, 26, 28. Similar sequences are obtained by starting with pairs other than 1 and 2. For example, if you start with 1 and 3 you get the sequence which begins 1, 3, 4, 5, 6, 8, 10, 12, 17, 21.

Using your micro, it is an easy task to generate such sequences to very great lengths. Then comes the hard part of the problem. By examining such sequences, what, if anything, can you say about them? As Ulam says on page 89 of his book, this question seems very hard to answer.

Abstract and irrelevant? In the exact form it is stated the above problem may indeed be just a number game. But, as your neighborhood physicist will tell you, it's a question which is awfully close to the kind of considerations that arise in nuclear interactions— in particular in atomic bombs. As Ulam knew only too well, everything in life is a game. His great success came from never forgetting that maxim. I wish I had the chance to meet him. But at least now I have this book.

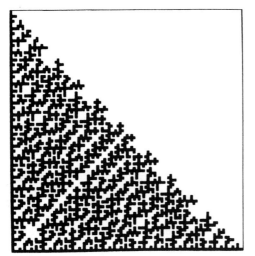 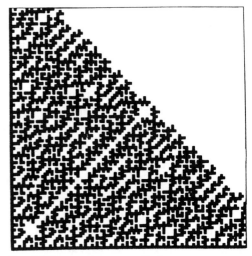

FIGURE 26.   Ulam patterns grown from a single square. 100 and 120 generations, respectively.

72

Can you crack the code of the spilled nail varnish?

July 24, 1986

This week's column is devoted to a collection of teasers to test both your brain and that of your computer. Answers are given at the end.

1. While playing with my daughter's calculator the other day, I found that she had spilled nail varnish on the display, so that only the final four digits were visible. On entering a certain 4-digit number and pressing the square button, the result was the same 4-digit number. By chance, I had entered the only 4-digit number which would have produced this result. What is the number?

2. If you multiply 21 by 87, the answer, 1827, consists of the same four digits that made up the original two numbers. How many more pairs of 2-digit numbers can you find with this property, assuming all four digits are different? This is a good question for computer attack.

3. What are the two smallest whole numbers, the difference of whose squares is a cube and the difference of whose cubes is a square?

4. Prime numbers often figure in this column. A prime number is one which has only two divisors, itself and one. At the other end of the scale, what is the number below 1,000 which has the greatest number of divisors?

5. Finally, a question to which no answer is known as far as I am aware. If you take any number, reverse its digits to form another number, and then add the two numbers, and

if you then repeat the procedure for that new number, you are likely to eventually arrive at a palindromic number, that is one which reads the same in either direction. For instance, start with the number 19, reverse it to give 91, add 19 and 91 to give 100, reverse it to get 001, add to give 101, and there you have a palindromic number.

For most numbers chosen at random the process produces a palindrome very quickly, but there are some exceptions. For instance, if you start with the number 89 it takes 24 steps to get to the palindrome 8,813,200,023,188. It has been suggested that, provided you are prepared to go on long enough, you will always get to a palindrome by this procedure, no matter what number you start with. An easily written computer program should help you to collect evidence for this conjecture. Until, that is, you come to look at the number 196. To my knowledge this has been run for at least 75 steps without hitting a palindrome, and I dare say it has been taken further.

But even if this number does turn out to give a palindrome, the general question of whether every number does still remains unanswered.

Answers

1. 9376.

2. There are another six such pairs: 15 and 93; 21 and 60; 21 and 87; 27 and 81; 30 and 51;

plus one further pair that you will have to find yourself.

3. 6 and 10;

$$10^2 - 6^2 = 4^3 \quad \text{and} \quad 10^3 - 6^3 = 28^2.$$

4. 840, which has 37 divisors.

Postscript. See Chapter 75 for an update on the palindrome problem.

73

Wallpapering by numbers

July 31, 1986

How would you like to design your own wallpaper so that no one else in the world would be likely to have the same pattern? Imagine a system where you go into the shop and key a couple of secret numbers of your own choice into a computer terminal, and a computer-based pattern designer creates your wallpaper pattern according to your chosen numbers.

Whether or not such a system already exists (if it does, it is probably in California) I do not know, but the mathematics for such a device certainly is available. In fact, there are a number of ways of setting up such a system, and anyone with a home micro can write a simple program which can be used to design an infinite variety of beautiful and symmetric patterns.

The pattern shown was obtained by Barry Martin, a mathematician at Aston University. Using a VAX 11/750 computer together with a Tektronix 4113 graph plotter, Martin has been able to generate many very detailed pictures of this type. An average home micro would take longer to generate its pictures, and the quality would not be so high, but apart from that there is nothing to stop anyone producing similar diagrams.

Martin became interested in this work last year when in the hospital. To while away the hours, he borrowed his son's 48K micro, got a porter to provide him with a black-and-white monitor (which, it later transpired, had been lifted from the operating theater) and got down to business. The basic idea that he followed was to generate the coordinates x, y of points in a simple iterative fashion.

Choose some function $f(x, y)$. The exact choice of function is up to you, but it should not be linear, i.e., it should involve squaring or cubing at least one of x and y, for instance the function $x^2 + xy$.

Then pick some point $x_0 y_0$ on the screen. (You will have to experiment with the various parameters involved to get the scale right.) Now generate an entire sequence of points according to the rule

$$x_{n+1} = y_n - f(x_n, y_n)$$
$$y_{n+1} = A - x_n$$

where A is some numerical parameter, which it is also up to you to choose.

Plot the points x_n, y_n as you go, and keep the iteration going until you have enough points to provide an acceptable picture. Some choices of the function $f(x, y)$ will not lead to anything very exciting, others will give you some extremely attractive patterns. Proceed by trial and error, as did Martin (who is keeping secret the functions he uses). The choice of the parameter A can also have a significant effect.

Colors may be used to enhance the patterns, for example by assigning color according to how far the point plotted is from the origin of coordinates, or else by means of some rotational system. All you need is a little patience while your patterns build up. Martin cautions that this can often take some time.

He has found patterns which seem to build up in one area of the screen for thousands of iterations, and then suddenly the whole process jumps to another region and a further piece of the pattern develops.

When you have found a function $f(x, y)$ which works well for different values of the parameter A, you can contemplate setting up in the customized wallpaper business.

Finally, here is a simple-sounding little problem that could have you wasting many hours of your time. What 10 figure number has the property that the first digit tells you the number of zeros in the number, the second digit tells you the number of 1s, the third digit the number of 2s, and so on up to the tenth digit which tells you the number of 9s? [See Chapter 76 for the solution to this problem.]

FIGURE 27.   Barry Martin's wallpaper: secret pattern.

74

Circling round the square

August 14, 1986

Circles and squares abound in architecture and design, as do the more general shapes of ellipses (symmetrical egg shapes) and rectangles. Both have, in their own way, pleasing symmetries that lead to useful functions. Is it possible to capture the best of both worlds by means of a shape that is midway between the circle and the square or, more generally, the rectangle and the ellipse?

During the late 1950s, the city planners in Stockholm, Sweden decided to completely rebuild the heart of the city. The design had to incorporate two major roads, one running north–south, the other east–west. The mathematical problem of the square-circle arose when it came to figuring out what to do where the two roads intersected at Sergel's Square.

What the planners wanted was some kind of a "square" (in the architectural sense, as in "Washington Square") with the roads running round it. But what geometrical shape was this "square" to take? A straightforward rectangle does not make for a smooth traffic flow, but a rectangle with rounded corners does not look very attractive. An ellipse might be aesthetically pleasing, but it tends to be too pointed to allow for good motoring.

Various combinations of circular arcs of different radii were looked at, but none was able to satisfy the famous Swedish passion for mathematical beauty. Driving in Stockholm,

the city planners agreed, should offer a beautiful mathematical experience.

So the planners took their problem to Piet Hein, a man whose talents are so varied as to render it impossible to describe exactly what it is he does for a living. To mathematicians and puzzle addicts he is the inventor of the game *Hex* and of the infuriating *Soma Cube* puzzle. To others he is known as one who writes on a variety of scientific and humanist topics. And still others see him as a prolific poet having a wry sense of humor.

The result of Hein's deliberations was the *superellipse* that eventually formed the basis of Sergel's Square. It was discovered using a computer to draw various sample curves of the same type, and choosing the one that seemed the most suitable. Anyone in possession of a home micro can generate a whole family of different superellipses with relative ease.

As every high school pupil should know, the equation for an ellipse whose longest radius is a and whose shortest radius is b is

$$\left|\frac{x}{a}\right|^2 + \left|\frac{y}{b}\right|^2 = 1$$

where the vertical lines in this formula denote the absolute value function, the result of ignoring any minus sign the number may have.

If a and b are equal, the equation produces a circle of radius a. If a and b are unequal, you get a genuine ellipse. What Hein did was investigate what happens when you modify the equation for an ellipse to use different exponents besides 2. That is, he looked at a whole range of curves produced by equations of the form

$$\left|\frac{x}{a}\right|^n + \left|\frac{y}{b}\right|^n = 1$$

for different (fractional) values of n.

For values of n less than 1, the equation produces a pleasing figure consisting of four concave curves. The closer n gets to 0, the sharper the bend in these four curves becomes, so that the figure resembles a four-pointed star. (For $n = 0$ you would just get the two coordinate axes, if your computer were smart enough to figure out what to do in this degenerate case.)

For n equal to 1, you get a straight-edged diamond shape.

For n greater than 1 you start to get ellipse-like shapes. At $n=2$ there is the usual ellipse. Then for n greater than 2, you get Hein's superellipses. The bigger n becomes, the more these begin to resemble a rectangle, and again a "smart" computer (or any competent math-ematician) could figure out that if it were possible to let n be infinity, the result would be a genuine rectangle measuring $2a$ by $2b$.

What value of the exponent n gives the superellipse that the eye finds the most pleasing? Hein thought that $2\frac{1}{2}$ was the optimum choice, and the Stockholm city planners agreed with him. For this was the shape they finally used in their design.

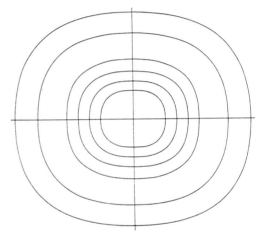

FIGURE 28. Concentric superellipses.

75

Back to front

August 28, 1986

The main topic this week is numerical palindromes.

There is obviously no interest to be had in simply looking for palindromic numbers, since it is a trivial matter to produce as many as you want. Palindromic dates provide slightly more interest. The next palindromic year is 1991. When is the next palindromic full date,* given that today is 28.8.1986?

Hunting for palindromic primes also provides the mathematically-minded computer enthusiast with some good raw meat to chew on. A great many have been found, and it has been conjectured that there are infinitely many such, though I am fairly certain this has not been proved absolutely. I am not sure what is the largest known palindromic prime. Perhaps readers with access to a supercomputer such as the Cray-XMP (which found the largest known prime of any kind) can enlighten me.

Any palindromic prime other than 11 will have an odd number of digits, since every palindromic number with an even number of digits is divisible by 11, and so cannot be prime.

Odd and even digit counts also figure in the problem of palindromic perfect squares. It is not hard to write a program to look for palindromic squares and if you do you will be able to discover very many, such as $121 = 11 \times 11$, $676 = 26 \times 26$, and so on. But practically all such palindromic squares have an odd number of digits. The first one with an even number of digits is 698,896, which is 836 squared. The next two are the squares of 798,644 and of 64,030,648.

When this column first began three years ago, these were the only ones known, but since then two regular readers have responded to my articles on the topic by using their micros to discover two more. In 1984, Graham Lyons of Romford in Essex ran his IBM-PC for an entire weekend to discover a 22-digit palindromic square, namely the square of 83,163,115,486. Then, in April of this year, Jeremy Stanforth, who lives in Munich in West Germany, ran his 48K home computer for 20 hours to find one with 26 digits. This result demonstrates that if you are lucky, even a modest home computer can yield results.

Stanforth started his search quite arbitrarily at the point 6,325 billion (a 13-digit number, whose square will have 26 digits), and was rewarded with a palindromic square when he reached the number 6,360,832,925,898. (Check this for yourself. If the square is not palindromic, you will know there has been a misprint, in which case you will have to get

---

* The date is given here in British fashion, with the day first, then the month, and finally the year.

your own computer working on the problem. At least you know roughly where to look!)

Another problem about palindromic numbers was stated in my article on July 24 [Chapter 72] I observed that if you take any number, reverse it and add the two numbers you now have, then do the same for the total, and keep on repeating this procedure of reversing and adding, then it is likely that you will eventually reach a palindromic number.

For example, starting with 38, reversing it gives 83, adding 38 and 83 gives 121, and after just one step you have reached a palindrome. Start with 19: reversing it gives 91 and adding 19 and 91 gives 110. This is not a palindrome. Reverse this number to get 011, then add to give 121, a palindrome (which happens to be the same as the one before—or is this more just an accident?).

It has been conjectured that no matter what number you start with (with as many digits as you like), the result will always be the same. Eventually a palindrome will result. But this has not been proved, and in fact a number has been discovered which seems to refute the conjecture. The number is 196, and in my previous article I stated that I knew it did not give a palindrome in under 75 steps.

Graham Lyons (mentioned above) and G.P. Davies of Ealing (using an Apricot PC running compiled BASIC) took the process up to 24,093 steps (where the number looked

at exceeds 10,000 digits) without encountering a palindrome. (Several other readers also joined in the hunt, but their micros gave up before reaching such dizzy heights).

Perhaps *Micromaths* will be able to claim another discovery at this rate. Newcomers who want to join the search can check their programs on the starting value 89, which after exactly 24 steps produces a 13-digit palindrome.

Two quick comments on former problems. My July 24 column included a request for a 4-digit number such that when you square it, the last four digits of the answer give the original number. The only answer is 9376. Several readers wrote in to suggest 0652 as well (including one reader who worked on the problem on a pocket calculator while on the summit of Snowdon, and yes, Mr. P. Reynolds, it was foggy when I was there too, but I forgot to take a calculator to pass the time).

But 0625 is really only a 3-digit number. If you allow leading zeros there are two further solutions, 0000 and 0001.

On July 31 [Chapter 73] I asked for a 10-digit number whose first digit gives the number of zeros in the number, whose second digit gives the number of ones, and so on up to the tenth digit which tells you the number of nines. There is only one such number, despite a number of claims to the contrary. I shall tell you it next time.

76

Seven-up*

September 11, 1986

The number seven is, without doubt, one of the most common numbers that occur outside of mathematics. Seven days in the week, seven classical planets (Moon, Mercury, Venus, Sun, Mars, Jupiter, and Saturn), seven notes on the musical scale, seventh sons of seventh sons, the seven deadly sins, the seven wonders of the world, seventh heaven, the seven Sacraments, the seven seas, and even the name of a popular soft drink.

As a mathematician, I cannot believe that there is anything to this other than wish fulfillment. Once word got around that it took seven days to make the Universe, trying to fit various phenomena into a framework involving the number 7 was practically unavoidable.

The Pythagoreans went to great lengths to find significance in the number, observing that a child could (only could, mind) be born seven months after conception, would cut its teeth seven months after birth, reach puberty after two periods of seven years, become an adult after seven more years and so on.

Even so, the number seven does have plenty of mathematical significance. One of the most intriguing properties of the number occur when you look at reciprocals of whole numbers. In the case of 7, if you work out 1/7 as a decimal you get the infinite repeating pattern:

$$1/7 = 0.142857\ 142857\ 142857\ldots\,,$$

where the sequence 142857 keeps repeating ad infinitum. The amazing thing about this is what happens when you multiply by any of the numbers 2, 3, 4, 5, 6 to obtain the decimal expansions of 2/7, 3/7, 4/7, 5/7, 6/7, respectively.

In each case, the result is exactly the same pattern of digits shifted along one or more places. For example,

$$3/7 = 0.42857\ 142857\ 142857.$$

Seven is the first number with this property. The next is 17, where you get an infinite repeating pattern of 16 digits which simply shifts along when you multiply by any of the numbers from 2 to 16. Other examples are 19, 23, 29, and 47. There are many more such numbers and it is not hard to write a computer program to find them, though it has not been proved that there are infinitely many such. All you need to do is write a routine to perform long division of any whole number into 1.

A number of readers sent in solutions to my problem of six weeks ago [Chapter 73] to find the only 10-digit number whose first digit

---

* The original title for this piece was "Come in number seven," a phrase having a well-known and amusing meaning in Britain at the time. I won't even try to explain it here.

tells the total number of 0's in the number, whose second digit gives the number of 1's, and so on up to the tenth digit which gives the number of 9's.

The answer is 6,210,001,000. There is a fairly easy way to discover this number. (Dave Healey of Faringdon in Oxfordshire, writes that his nine-year-old son used the method successfully.) Start with any good "guess" at the answer—for example, the number 9,000,000,000. This misses the required property only because there is one nine in the number and yet the last digit is zero, not a one. Starting with this number, go through it digit by digit, changing each digit to its "count". This gives you the number 9,000,000,001. Now do the same with that number to get 8,100,000,010. Then again, to get 7,210,000,100. The next step gives the answer. Try this for yourself with other starting values, for instance 9,876,543,210.

A related problem is to see if you can find any numbers with fewer than ten digits with similar properties, such as 9-digit number using the digits from 0 to 8, and so on. All these problems lend themselves to computer investigation. (For instance, Dave Healey, mentioned above—who tells me that he is not a mathematician—programmed his computer to see how many steps it took to get the answer to the 10-digit problem starting from different initial guesses. In many cases four or five steps are enough.)

When you have sorted out the above, try looking for the 9-digit number which, when multiplied by the number 123,456,789 gives an answer whose last nine digits consist of the number 987,654,321. For those of you who give up, I shall give the answer next time [Chapter 77].

77

Pi in the sky

September 25, 1986

It is a pity that it was not done using a Macintosh computer. Then this article could have carried the headline "American Apple Pi." Unfortunately, powerful though the Mac is as a business micro, it just ain't up to calculating the decimal expansion of the mathematical constant π to record numbers of places. For some years now this has been the exclusive domain of all-powerful supercomputers.

The latest group to get in on the act is the NASA Ames Research Center in Mountain View, California, who recently used their newly installed Cray-2 computer system (the latest in a short but highly significant line of high-performance mainframe computers produced by the Cray company) to calculate π to a startling 29,360,000 decimal places.

As every schoolchild knows, the number π is what you get if you divide the circumference of any circle by its diameter. Commonly used values for π are $22/7$, or the more accurate 3.14159, but both of these are mere approximations to the true value. In fact, it has long been known that there is no way of calculating the exact value of π. It is what mathematicians call an "irrational number." That means that its decimal expansion would (if it could be calculated) involve infinitely many places, and moreover the digits appearing in these infinitely many places would do so in a seemingly random fashion, unlike "rational" numbers such as the fraction $1/7$ discussed two weeks ago [Chapter 76], which gives the infinite decimal expansion $0.142857\ 142857\ldots$, where the same six digits 142857 keep on appearing ad infinitum. No such pattern occurs with π, whose decimal expression begins $3.14159\ 26535$ and then keeps on going in an apparently haphazard fashion.

Methods for calculating π to various degrees of accuracy have been developed since ancient times. A Babylonian clay tablet from around 2200 B.C. that was discovered in 1936 and translated in 1950 gives the value 3.125, while an Egyptian papyrus roll from around 2000 B.C. gives the value 3.1605. Both of these are better than the value of 3.0 that is implicit in the bible (I Kings 7:23 and II Chronicles 4:2). The value of $22/7$ mentioned earlier was discovered by the Greeks around 250 B.C.: A better fractional approximation is the one $355/113$ discovered by the Chinese astronomer Ching-Chih.

In more recent times, a German called Ludolph von Ceulen spent a large part of his life calculating the wretched number to 35 places and had the result engraved on his tombstone, and achieved some sort of immortality in that even to this day some German authors still refer to π as the "Ludolphian number."

In 1699, Sharp went up to 71 places, and in 1706 Machin succeeded in getting up to 100. Then followed 200 places by Dase in 1824, 500 places by Richter in 1854, and 707 places by

William Shanks in 1874. Fortunately, Shanks was long gone when desk calculators were brought to bear on the task in 1945, when a mistake was discovered in the 527th place, thereby rendering as totally useless a lifetime's work that would otherwise have been merely virtually useless.

Modern computers have, of course, made much greater tallies possible. In 1973, Guilloud and Bouyer of France went up to one million places, and published the result in a book. In 1981, after a run of 137 hours on a Facom M200 computer, a Japanese mathematician, Kazunori Miyoshi, obtained a printout of two million places.

Two years later two more Japanese, Yoshiaki Tamura and Yasumasa Kanada, used a Hitac M-280H computer to calculate 10 million places. But though the same pair eventually went beyond this total, it was clearly only a matter of time before good old American "can do" took over.

"Why did they do it?" you may ask, not unreasonably given that the highly expensive Cray-2 system that NASA bought was presumably obtained for more significant tasks, such as getting the Space Shuttle airborne again. The answer is provided by the man responsible, Sterling Software's David Bailey, a consultant with the Numerical Aerodynamic Simulation Program at NASA.

The calculation was, he says, just what was needed to test the accuracy and efficiency of the newly installed computer. For such is the nature of the computation that a single error in just one place rapidly gives rise to a completely meaningless answer. So by performing the calculation twice using different programs (involving different mathematical ways of getting at π) and comparing the two answers, a highly reliable performance check results. To be precise, a check that incorporates over 30 trillion arithmetic operations, utilizing over 100 million words of main memory, running for almost 30 hours. This ought to give some measure of comfort to future Shuttle crews, as well as getting NASA into the *Guinness Book of Records*.

In answer to the problem posed in my last column [Chapter 76], the number which when multiplied by 123,456,789 gives an answer ending in 987,654,321 is 989,010,989.

Postscript.   See Chapters 129 and 131 for an update on the calculation of π.

78

Friendly numbers
October 9, 1986

Amsterdam-based mathematician Herman J. J. te Riele has just announced that he has discovered a method that enables him to compute all pairs of "amicable" numbers up to any pre-set limit. Using his new method—which has been awaiting discovery for more than 2,000 years—he has catalogued all amicable pairs whose smallest member is less than 10 billion.

The entire list of the 1,427 amicable pairs in this range is reproduced as a 30-page supplement to his paper in the latest edition of the journal *Mathematics of Computation,* Vol. 47, No. 175, pp. 361–368 (supplement on pp. S9–S40). The calculation of this list took over 1,000 hours of computer time on a Cyber 750. More than 800 of the pairs listed had not been discovered before the arrival of the new method.

So what, exactly, is an amicable pair of numbers? The idea dates back to at least 500 B.C., and was certainly known to the Pythagoreans in Ancient Greece. The smallest amicable pair consists of the two numbers 220 and 284. If you list all the proper exact divisors of 220 you get a list of 11 numbers, namely 1, 2, 4, 5, 10, 11, 20, 22, 44, 55, 110. Add all of these together and you get 284. Now take the number 284 and add together all its proper exact divisors. What do you get? You get the number 220 again. And that is what an amicable pair is. Two numbers with the prop-erty that each is the sum of the divisors of the other.

The concept of an amicable pair is obviously a derivation of that of a perfect number (one which is itself equal to the sum of its divisors, such as 6 or 28).

Just as the perfect numbers have attracted a fair amount of somewhat dubious mystical speculation, so too have the amicable pairs come in for their share of canonization. For instance, in the Bible (Genesis 32:14) we read of Jacob's present of 200 she-goats and 20 he-goats to his brother Esau. The fact that the total number of goats involved here, 220, is one member of the amicable pair mentioned earlier, has led at least one Bible commentator to refer to its significance in Jacob's trying to secure Esau's friendship. (Of the assorted 30 milch camels and colts, 40 kine, ten bulls, 20 she-asses, and ten foals that went with the goats, I have not come across any mathematical significance. But doubtless readers will inform me of any such.)

Pythagoras himself said that a friend is "one who is the other I such as are 220 and 284," while the 11th century Arab El Madschriti managed to anticipate a future time when certain British tabloid newspapers might include a mathematics column by claiming that amicable numbers possessed erotic powers, the idea being that the eager male of the species should eat the number 284 (in the form of a decorated cake, one presumes) while giving the object of his desire the number 220.

Though I have to confess that I have so far resisted all temptation to try this novel approach, past experience does lead me to speculate that any attempt to introduce mathematics into one's amorous activities is likely to meet with very little success.

Over the centuries, a number of mathematicians, both great and otherwise, have tried to find new amicable pairs of numbers. The famous Swiss mathematician Leonhard Euler listed 60 in 1750, though, curiously, both he and everyone else missed the pair 1,184 and 1,210, the second smallest pair, which was discovered as late as 1866 by a 16-year-old boy, B. N. I. Paganini. Various methods were worked out for finding amicable pairs systematically, though until the recent te Riele result none of them gave all possible pairs.

The more recent work has all made heavy use of computers in order to handle the large numbers that are encountered. It is still not known if the list of amicable pairs continues indefinitely or if there is a largest such pair, though most experts are inclined to guess that there are infinitely many of them.

Using a microcomputer to try to find amicable pairs by a simple search (using a routine that lists and adds together all the divisors of a given number) is not particularly entertaining, since such pairs are rare. What can be done at home, without any higher mathematical theory to go on, is this: take any number; add all its divisors to get a new number; now add all the divisors of that new number to get a third number, and so on.

Most of the time you will eventually end up with number 1. But sometimes the divisor sums get bigger indefinitely. You might even hit upon what is known as an amicable chain, where after a certain number of iterations of the divisor-adding process you find you are back to your original number. For example, starting with the number 14,316, you discover that after 28 steps you get back to that number again. Investigation of this simple concept can lead to various further computer experiments and might even provoke you into putting forward (and then testing on the computer) your own theories about such behavior.

79

New life for good old numbers

November 6, 1986

"When Dr. Johnson felt, or fancied he felt, his fancy disordered, his constant recurrence was to the study of arithmetic." So wrote James Boswell in his celebrated *Life of Johnson* in 1791.

Johnson himself, quoted in the biography, says: "All minds are equally capable of attaining the science of numbers: yet we find a prodigious difference in the powers of different men, in that respect, after they have grown up, because their minds have been more or less exercised in it."

Many famous mathematicians, of course, were keen on arithmetic. Men such as Gauss, Fermat and Euler, who produced masses of highly abstract mathematics with hardly a two-figure number in sight, all spent a lot of time working (they probably called it playing) with such common or garden objects as the positive whole numbers everyone learns as a toddler.

Nowadays, of course, arithmetic is completely out of vogue, and a generation has grown up without that familiarity with numbers that older people take for granted. I recently amazed a young supermarket cashier by saying that the total shown on the cash register could not possibly be correct as it was an even number: there was no way it could arise from my having purchased an odd number of cans of dog food all priced at 29 pence. After an inordinately-lengthy re-entering of the

items in the register she eventually arrived at the correct total, but was not able to see why I knew the first answer had to be wrong.

For her, as for most of her contemporaries, numbers were just things you punched into keyboards. She had no feeling for them. The computer, and before that the pocket calculator, had kept her and millions like her from ever getting to know numbers in the almost intimate manner of past generations.

But far from preventing anyone from gaining familiarity with numbers, the computer can, if used properly, help to create that awareness, while providing some freedom from the drudgery that any lengthy calculation entails. The diagram accompanying this article gives one example. All you need is a pocket calculator, or any home micro, to discover all kinds of similar patterns that are produced by basic arithmetic.

$$
\begin{aligned}
1 \cdot 1 &= 1 \\
11 \cdot 11 &= 121 \\
111 \cdot 111 &= 12321 \\
1111 \cdot 1111 &= 1234321 \\
11111 \cdot 11111 &= 123454321 \\
111111 \cdot 111111 &= 12345654321 \\
1111111 \cdot 1111111 &= 1234567654321 \\
11111111 \cdot 11111111 &= 123456787654321 \\
111111111 \cdot 111111111 &= 12345678987654321
\end{aligned}
$$

FIGURE 29. **Ones for the tree.** The hidden beauty of multiplication.

For instance, see what you get when you construct a table of all the products of 7 × 7, 67 × 67, 667 × 667, 6667 × 6667, and so on, where each multiplication is that of a number consisting of a row of sixes followed by a seven, with itself. Then try it with the squares of the numbers 4, 34, 334, 3334, etc. How about the numbers 9, 99, 999, 9999, and so on? Or if you want to see something different, multiply the "magic" number 15873 by each of the numbers 7, 14, 21, 28 (i.e., the multiples of 7). There are all sorts of patterns to be unearthed.

When you do find one, there is always the question, "Why does it happen?" This is just what the old masters like Fermat did all the time, and look where it got them!

But is this any use in today's high-tech world? Well, besides amazing supermarket cashiers, you can always use simple arithmetic to entertain children. For instance, think of any three-figure number. Now write the same number backwards and subtract the smaller of your two numbers from the larger. Take this new number, reverse its digits and this time add the two numbers. You will end with the number 1089.

Finally, what is the smallest whole number N with the property that if you move its first digit to the end the resulting number is exactly half as big again as N? I should warn computer buffs that this is an example where brain power is far more important than megabytes of silicon. I'll give you the answer next time [Chapter 80].

80

Valiant strides at the games

November 20, 1986

If you had been in the vicinity of the University of California at Berkeley between August 3 and 11 this year, you might have been treated to the spectacle of earnest-looking men and women poring over scribbled hieroglyphics on table napkins and discussing matters such as the way in which compact simply connected topological four-dimensional manifolds are completely determined by their two-dimensional homology group H_2, the intersection form on $H_2 \times H_2$, and the Kirby–Siebenmann obstruction to stable triangulation (all of which, as any schoolboy knows, leads rapidly to a solution of the four-dimensional Poincaré Conjecture).

The occasion was the 20th International Congress of Mathematicians, a four yearly Olympic Games of mathematicians held at various locations around the world. Four thousand mathematicians from around the globe turned up for this year's binge, one of the highlights of which is always the awarding of the Fields Medals, the mathematicians' equivalent of the Nobel Prize. The obscure-sounding result quoted above resulted in its discoverer, Michael Freedman, receiving one of the three medals handed out this time.

The International Congress provides an indicator of the way mathematics is developing. The enormous changes brought about by computing and the challenges it raises for the mathematician were recognized by the intro-duction at the previous Congress of a special prize for research relating to information science: the Nevanlinna prize.

This year the award went to the 37-year-old Leslie Valiant, a British citizen of Hungarian origin, currently at Harvard University. His work has been in the subject known as complexity theory, which deals with the design of efficient computer programs. One of Valiant's results concerns the speed with which a computer can compile a program written in a high-level language such as Pascal or Fortran.

The first step in such a process requires the parsing or understanding of the instructions that make up the program. With traditional parsing techniques, parsing a program consisting of N instructions required a computing time proportional to N^3.

Valiant found a way of doing the job which requires a time proportional to only $N^{2.5}$. For large programs (where N can be into the thousands) this represents a considerable potential saving in expensive computing time, and is one more illustration of how investment in mathematical research can pay huge dividends in computing.

And so to a table of simple arithmetic sums which gives an attractive numerical pattern. Those who saw my last column two weeks ago [Chapter 79] will see how this follows on from what was discussed there. The table given then involved the squares of the numbers consisting only of ones, namely 1, 11, 111, 1111, etc. Similar to the one given this time, see what happens when you work out the sums consisting of the numbers 9, 98, 987, down to 98765432, each multiplied by 9, and then add on 7, 6, 5, down to 0, respectively. Examples such as these provide a marvellous way of developing an interest in numbers among children, in an age when the computer and the pocket calculator have taken away the need for arithmetic skills in many aspects of everyday life.

Another nice result is obtained when you take the square of the number 1,111,111 and subtract 10 times the square of 111,111. When you have seen what happens in this case, try to explain why you get the answer you do. If you manage this part, you are well on the way to becoming a good computer scientist, regardless of whether or not you can tell a hash table from a fast food counter.

$$1 \cdot 8 + 1 = 9$$
$$12 \cdot 8 + 2 = 98$$
$$123 \cdot 8 + 3 = 987$$
$$1234 \cdot 8 + 4 = 9876$$
$$12345 \cdot 8 + 5 = 98765$$
$$123456 \cdot 8 + 6 = 987654$$
$$1234567 \cdot 8 + 7 = 9876543$$
$$12345678 \cdot 8 + 8 = 98765432$$
$$123456789 \cdot 8 + 9 = 987654321$$

FIGURE 30.   Eight and counting. A curious number pattern.

Finally, last time I asked you for the smallest whole number N with the property that if you move its first digit to the end the resulting number is exactly half as big again as N? I warned you that this is an example where brain power is far more important than megabytes of silicon. Here is the answer.

The idea is to start out by letting $a, b, c, \ldots x, y, z$ be the digits of N in that order. (This is not meant to suggest that there are exactly 26 digits. There may be more or less. It just provides something to work with.)

Now look at the fractional number whose decimal expression is

$$a.bcd \ldots xyza\, bcd \ldots xyza \ldots$$

(etc, ad infinitum). That is, the digit a comes before the decimal point, then comes an infinitely repeating cycle of the digits $bcd \ldots xyza$. Call this fractional number G. If you subtract the single-digit number a from G you get the number having zero before the decimal point and then the repeating cycle $bcd \ldots xyza$. The property of N that is being considered tells you that $bcd \ldots xyza = \frac{3}{2}(abc \ldots xyz)$. So your number $G - a$ will in fact be $3/2 \times G/10$. Rearranging this in the form of an equation, you get $17G = 20a$.

The question asks for the smallest possible N, so G will also have to be the smallest possible, and this means that a has to be 1. So $G = 20/17$. In other words, the number N is the 16-digit monster 1,176,470,588,235,294 that you get by working out G as a decimal.

81

Living at the margin

December 18, 1986

Some time around 1637, the French mathematician Pierre de Fermat wrote a scribbled note in the margin of a mathematics book he was working from, and with that one simple act ensured for himself immortality in the minds of ordinary men. Were it not for that marginal note, there is little doubt that Fermat's fame would be among professional mathematicians alone. As it is, Fermat's Last Theorem (as the note is now known) is the most famous mathematical problem outside of the field of mathematics itself.

The fame arises through a combination of circumstances. For one thing, Fermat was one of the greatest mathematicians of all time, in spite of the fact that he regarded mathematics as nothing more than a hobby, being by profession a lawyer. For another, he published virtually nothing, relying on the communication of his discoveries in letters to other mathematicians to ensure their dissemination—usually leaving it to the recipients of his letters to supply the proofs of his claims. Since it almost always turned out to be the case that his claims were correct, it is clear that he usually knew what he was talking about.

This gives considerable weight to the note he scribbled in the margin of that book. It concerns the solution of equations of the form $x^n + y^n = z^n$, for whole numbers n, where the solution values for the unknowns $x, y,$ and z have to be whole numbers. The simplest

case (other than $n = 1$, which is trivial) is where $n = 2$. In this case the equation becomes $x^2 + y^2 = z^2$, and solutions are known as Pythagorean triples because of the obvious connection with Pythagoras' Theorem for right-angled triangles.

Two possible solutions are $x = 3, y = 4,$ $z = 5$, and $x = 5, y = 12, z = 13$. There are infinitely many others. Fermat asked himself about possible solutions to equations where the power n is bigger than 2. His marginal note claimed that whenever n was bigger than 2, the corresponding equation has no solutions at all (among the whole numbers). "For this I have discovered a truly wonderful proof," he went on, "but the margin is too small to contain it."

Was Fermat correct in his claim? Had he really found a "truly wonderful proof"? Or was the whole thing a well thought out practical joke on his successors? If it was, it turned out to be tremendously successful, since three hundred years effort by the greatest mathematical minds have so far failed to discover whether Fermat's claim is true or false. The closest that anyone has come to proving the result was the discovery three years ago by a West German mathematician, that for no number n beyond 2 does the equation $x^n + y^n = z^n$ have infinitely many solutions (except in the trivial way whereby any one solution can be multiplied up to give others). [See

Chapter 7.] Apart from this and some highly technical results, the problem set by Fermat's marginal note remains unsolved.

But what of that book in which the original note was made? Well, it was a copy of a translation of a work written by a certain Diophantus, who lived in Alexandria some time around A.D. 250. This book, called the *Arithmeticæ,* may be regarded as the first ever textbook on algebra.

It is in this book that there appeared the first account of how to set about solving equations in two or more unknowns, where a solution has to be found involving whole numbers only. For instance, one of the problems given in the *Arithmeticæ* is this. Find four numbers, the sum of every arrangement three at a time being given; say 22, 24, 27, and 20. (Book I, Problem 17). Readers should not have too much trouble with this one. All that is required is a little insight. (Failing that, a crude computer search would also work, but in this case that would really be doing it the hard way.) Nowadays, such problems are known as *Diophantine problems,* and an equation which has to be solved using whole numbers is called a *Diophantine equation.*

Obviously, because they involve whole numbers, Diophantine problems provide wonderful opportunities for the keen computer user to get his or her silicon teeth into. For instance, take a look at the equation $x^2 = 2y^4 - 1$. (That is, x-squared equals two times y to the power four, minus one). One solution (among whole numbers) is obviously $x = 1$, $y = 1$. But there is another solution. What is it? This is one where the computer user will usually outwit the pencil and paper fan. (Won't he?) What is curious about this particular problem is that there are just two answers. That this is so was proved conclusively in 1942 by the Norwegian mathematician W. Ljunggren. In general, all that can be said is that a Diophantine equation in two unknowns that involves one of those unknowns being raised to a power of 3 or more, can have only a finite number of solutions.

Many Diophantine equations turn out to have no solution at all. Like the Fermat equations perhaps? Computers are unlikely to help you there. Their use has shown that Fermat was certainly correct for all equations where the power n is less than 125,000. Beyond that no one knows, and the equation then involves quantities much too big to handle.

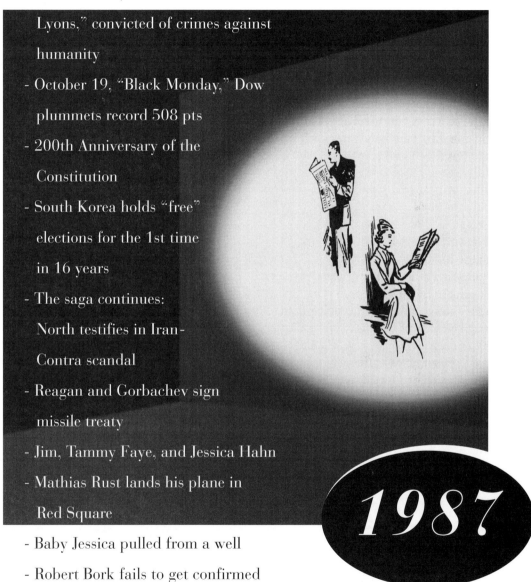

- *Time's* Man of the Year: Mikhail Gorbachev
- Klaus Barbie, the "Butcher of Lyons," convicted of crimes against humanity
- October 19, "Black Monday," Dow plummets record 508 pts
- 200th Anniversary of the Constitution
- South Korea holds "free" elections for the 1st time in 16 years
- The saga continues: North testifies in Iran-Contra scandal
- Reagan and Gorbachev sign missile treaty
- Jim, Tammy Faye, and Jessica Hahn
- Mathias Rust lands his plane in Red Square
- Baby Jessica pulled from a well
- Robert Bork fails to get confirmed
- *Thirtysomething*

1987

82

The art of the solvable

January 8, 1987

It helped put a man on the moon, and no doubt when a cure is discovered for the common cold, it will play a major role in that as well. The computer, that is. With these two standard debunking ploys out of action, it might begin to seem as though there is nothing the silicon marvel cannot achieve.

Each new year brings new advances both in the construction and use of computers. But you should avoid being taken in by all that skillful PR. There are plenty of things that computers will never be able to do, even in the field of mathematics. And I mean never. There are mathematical problems which have been demonstrated conclusively to be unsolvable by means of a computer, no matter how fancy and sophisticated they become.

One of the most illustrious of these totally impossible problems goes back to the turn of the century. In August 1900, the world's best mathematicians gathered in Paris for the second International Congress of Mathematicians, and among them was David Hilbert, the 38-year-old Professor of Mathematics of the University of Göttingen in Germany. As one of the leading mathematicians of the time (indeed of all time), Hilbert was due to give one of the keynote addresses of the meeting.

Hilbert chose not to follow the usual format and speak about recent developments in the field, but rather gave a list of what he regarded as the most significant open problems that would face mathematicians during the coming century. There were 23 problems in all. The tenth concerned a topic which went back at least as far as A.D. 250.

At that time, an Alexandrian mathematician by the name of Diophantus wrote a major textbook on the solution of algebraic equations among the whole numbers. This is the topic of Hilbert's tenth problem.

Solving polynomial equations among the real or complex numbers is a task most children learn to accomplish at school—at least as far as linear or quadratic polynomials in one unknown are concerned. Solving the equation

$$3x + 4 = 0$$

is straightforward, and there is a formula for solving quadratic equations like

$$x^2 - 4x + 5 = 0.$$

But when the solution has to be a whole number, things become less simple. If there are two or more unknowns, various outcomes are possible.

For instance, the equation

$$x^2 + y^2 = 2$$

has infinitely many solutions among real numbers, but only four solutions that are whole numbers (obtained by distributing plus and minus signs over the numbers $x = 1$, $y = 1$ in all possible ways). If the 2 on the right-hand side of this equation were a three, there would be no whole number solutions at all.

The following equation has just two whole number solutions

$$x^2 = 2y^4 - 1.$$

One solution is obvious: $x = 1, y = 1$. A computer search yields the other answer fairly easily. It is $x = 239, y = 13$.

As the above example should indicate, equations which involve whole numbers only are ideal for computer attack. But before you set your computer looking for a solution, it is probably wise to check that there are solutions to be found. Can this be decided by computer as well?

This is essentially what Hilbert asked. Is it possible to write a computer program that, when presented with any equation with whole number coefficients, can say whether or not there is a whole number solution? Of course, in 1900, there were no computers as we think

of them, so this is not exactly how Hilbert formulated his question. Rather he asked if there was an "algorithm"—i.e., a mechanistic, step-by-step procedure—that would do the job, and he would have been satisfied with a procedure even though it would be humanly impossible to ever build a machine to implement it.

The answer to Hilbert's tenth problem was provided only in 1970, by the 22-year-old Russian mathematician Yuri Matijasevič. No such procedure is possible. No matter how much computers develop in the future, it will never ever be the case that a program can be written that can decide whether a given whole number equation has a solution or not.

A particularly nice twist to the story is that Matijasevič's answer to the problem depended upon some earlier work of Alan Turing, the British logician whose work led to the construction of the modern computer as we know it.

83

Lies, damned lies, and logic

January 22, 1987

"Right teams, fingers on the button and here is your next question. Am I telling the truth or not when I say to you 'I am lying'?"

Buzz.

"Yes, Team A, What is your answer?"

"Well, Angela,* we think you're lying."

"Sorry, Team A. You see, if I were lying, then it would not be true that I was lying, so in fact I would be telling the truth. Team B, it's all yours now."

"We think you're telling the truth."

"Sorry, Team B. If I were telling the truth, then it would be true that I was lying, so in fact I would be lying."

"But Angela . . ."

"Sorry, teams, that's all we have time for this week."

When you have come to terms with that little conundrum, take a look at this one. If I say to you "I am telling the truth," am I in fact telling the truth or not? Well, if I am, then what I say will be true, so I will in fact be telling the truth. On the other hand, if I am not telling the truth, then what I say will be false, so I will not be telling the truth. Which seems to amount to . . . just what? Well, that I am both telling the truth and not telling the truth. Or at least, that either alternative is equally possible.

The first of these two puzzles is known as the Liar Paradox: a sentence that can be neither true nor false. The second is called the Truth-Teller Paradox: a sentence which can be either true or false, seemingly at whim.

Both have plagued logicians for centuries. The difficulty is that these sentences are so simple that one feels intuitively that each ought to be either true or false outright.

Or how about this one, known as Loeb's Paradox? Start out with the sentence that says: "If this sentence is true, then Chelsea** will be League champions." Call this sentence A. Suppose you wanted to see if A were true. You would start out by assuming that the antecedent in A was true and then try to deduce the consequent. If you could do this, then A would be true. So, assume the antecedent is true. That is, assume the phrase "This sentence is true" is true. In other words, assume A is true. Then, since A is true, the implication it expresses must be true. But, by assumption, the antecedent of that implication is true. So the conclusion of A will be true: that is, it will be true that Chelsea will be League champi-

---

*The most popular television quiz show host at the time was a former TV news anchorwoman called Angela.

**An English soccer club, based in London. At the time of writing this article, Chelsea was lying at the bottom of its soccer league.

ons. So you will have verified that sentence *A* is true (i.e., expresses a true implication). Since *A* is true, the antecedent of *A* is true, of course. (Read what it says). But if *A* is true (i.e., expresses a valid implication) and if the antecedent in *A* is true, then the consequent in *A* has to be true. So it really must be the case that Chelsea will be League champions. The logic seems to be inescapable. In spite of the fact that at the time of writing this article, Chelsea is, it happens, wallowing at the bottom of the First Division.

Paradoxes such as this seem to achieve their effect by some form of self-reference. Each of these paradoxical sentences speaks about itself. But there is nothing intrinsically wrong in allowing sentences to refer to themselves. Indeed, such self-reference can often convey a valid point. This sentence is a good example! So the problem facing the logician or the linguist is to try to work out just what exactly gives rise to the paradox.

Some interesting work of the Stanford-based logicians Jon Barwise and John Etchemendy has thrown considerable light on this vexing problem. And one of the most interesting aspects of the whole affair is that their work depends upon developments in computer science. So the machine orientated subject that first grew out of mathematical logic has finally started to give something back.

Several strands of computer science research went into the Barwise and Etchemendy results. Working at the Stanford University *Center for the Study of Language and Information,* Barwise and others had been working on a new mathematical theory to underpin machine translation and the design of Artificial Intelligence systems. Etchemendy, also at CSLI, was meanwhile trying to make sense of paradoxes such as The Liar, and seeing if the work of Barwise could shed any light on the problem. Then the Manchester-based logician Peter Aczel spent some moths in Stanford trying to develop a mathematical theory to handle computations that involved a degree of circularity (self-reference, if you like). Put the three together and what do you get? A resolution of the paradoxes: or at least, an explanation that goes some way to highlighting what exactly is going on.

In terms of the Barwise and Etchemendy framework, the Liar Paradox, for instance, is explained as follows. (I am simplifying considerably at this point.) Imagine a huge database containing all the facts of the world. Now ask your database if the Liar sentence is stored in the database (i.e., is true) or not (i.e., is false). The computer will search through its memory and will not find the Liar sentence among the true facts. So it will inform you that the sentence is not there. But here is the rub. If you now ask the computer if the fact that the Liar sentence is false, is a stored (i.e., true) fact, it will not be able to find that fact either. So the Liar sentence will be false (i.e., not true) but the fact that it is false will not be a registered fact.

Or, to put it another way, you will know that the Liar sentence is not true, but it will not be a fact that it is false (in the sense of your database).

In other words, what is at issue in the Liar Paradox seems to be not so much what is true and what is false, but *what can be known about* the truth or falsity. Computers would have no trouble with the Liar sentence. Only humans. Or so it seems. (Do you believe me?) In any event, it is nice to see work in computer science helping to resolve some ancient puzzles about language that go back to the time of the Ancient Greeks.

84

Rabbit pi*

February 5, 1987

Rabbit pi. Familiar in the dining room, perhaps, but in mathematics? Well, it is now, thanks to a recent result of the Soviet mathematician Yuri Matijasevič, which shows that there is an intimate (though sophisticated) connection between an old problem of Fibonacci concerning rabbits, and the mathematical constant π.

The so-called Fibonacci sequence gets its name from the 13th century Italian mathematician Leonardo of Pisa, who wrote under the name of Fibonacci (from the Latin *Filius Bonacci,* meaning of Bonacci). It was Fibonacci's book *Liber Abaci* that introduced into Europe the powerful Hindu-Arabic decimal notation that we use today. It also presented mankind with the following problem.

A man puts one pair of rabbits in a certain place surrounded by a wall. How many pairs of rabbits can be produced from that pair in a year, if the nature of these rabbits is such that every month each pair bears a new pair which from the second month on becomes productive?

It does not take long to figure out that the number of pairs of rabbits present each month is given by the sequence 1, 2, 3, 5, 13, 21, 34, ... Turning the problem into a piece of general mathematics, the Fibonacci sequence is defined by the iterative formula

$$u(0) = 0 \qquad u(1) = 1$$
$$u(n + 1) = u(n) + u(n - 1)$$

So $u(2) = 1$, $u(3) = 2$, $u(4) = 3$, and so on, giving the rabbits sequence.

The Fibonacci sequence crops up in many parts of mathematics and computer science: for instance in the study of algorithm efficiency and in database design. But it is not at all apparent that it is connected with that ubiquitous mathematical constant π, the number obtained by dividing the circumference of any circle by its diameter.

But is there indeed a connection between π and the Fibonacci sequence? Suppose you start out with the Fibonacci numbers $u(1)$, $u(2)$, $u(3)$, etc., up to some point $u(m)$. Multiply them all together to give a single number, $P(m)$. When you have done that, calculate the least common multiple $\text{lcm}(m)$ of those same Fibonacci numbers $u(1)$ through to $u(m)$. Then work out the quotient $6 \log P(m) / \log \text{lcm}(m)$.

Now take the square root of this quantity, call it $Z(m)$. Then, if you were to do this for increasing values of m, the numbers $Z(m)$ you got would get closer and closer to π. By

*This is a shortened version of an original article entitled "Pulling pi out of the hat."

calculating the number $Z(m)$ for sufficiently large m you can calculate π to however many decimal places you require—in principle, at least. However, this is not intended as a new method for calculating π. Rather it demonstrates that, however unlikely it may seem that Fibonacci's rabbits ever got near to ending up in a pie, and however different the Fibonacci sequence may seem mathematically from π, the two notions are nevertheless connected. And that is what is surprising.

For those giving the method a try, the value $m = 7$ gives

$$P(m) = 1 \times 1 \times 3 \times 5 \times 8 \times 13 \times 21 = 32760$$

$$\text{lcm}(m) = 10920.$$

Taking logs to base 10 (it does not matter which base),

$$6 \log 32760 / \log 10920 = 6 \times 4.515/4.038$$

$$= 6.709.$$

Taking the square root gives $Z(m) = 2.590$. This is not very close to π yet, but why don't you see what happens for significantly larger values of m. Using a computer, it's as easy as pie.

Reference

The American Mathematical Monthly, 93 (October 1986) pp. 631–635.

85

Prime revelations

February 19, 1987

Smith numbers first emerged about five years ago as a result of an observation by Professor Albert Wilansky of Lehigh University in Bethlehem, Pennsylvania. What he observed concerned his brother-in-law's telephone number: 493-7775. Since the brother-in-law was called Smith, the origin of the term "Smith number" is clear. But just what is it that is so special about this telephone number, and what exactly is a Smith number?

It's all a question of the way the number splits up into a product of prime numbers. A prime number is a whole number such as 2, 3, 5, 7, 11, 13, 17 that cannot be split up into a product of two or more smaller numbers. Every whole number is either a prime number or else splits up (in an essentially unique way) into a product of prime numbers.

It is a straightforward task to write computer routines that test a given number to see if it is a prime and to split up any non-prime into a product of primes, and in this way it is possible to produce a table of prime numbers and factorizations of non-primes into products of primes.

If you do this, then amongst the entries you will find the following: $4 = 2 \times 2$, $22 = 2 \times 11$, $27 = 3 \times 3 \times 3$. In each of these cases, if you add up the digits in the number you will get the same result obtained by adding together all the digits in the prime factorization. So, in the case of 27, adding together 2 and 7 gives 9

and adding together the three 3s also gives 9. Likewise with Mr. Smith's telephone number. The prime factorization of the number 493-7775 is $3 \times 5 \times 5 \times 65837$. Add together the digits in the telephone number and you get the answer 42. Add together the digits in the prime factorization and you also get 42.

And that is what a Smith number is: one where the sum of the digits is equal to the sum of the digits in the prime factorization of the number.

Smith numbers are not very common. Wilansky himself checked all numbers less than 1,000 and found that only 47 of them are Smiths.

In 1983, two characters by the name of Oltikar and Wayland found examples of Smith numbers much larger than those of Wilansky, the biggest one being a number with 362 digits. This was already far larger than the kind of number which can be split into prime factors using present techniques and computing equipment, for which numbers in excess of about 85 digits are too big to handle with any certainty of success. But, of course, in looking for Smith numbers there is no need to start with some large number and then try to factorize it. You can start with a collection of large prime numbers and multiply them together. This is what Oltikar and Wayland did. Their idea was to start with a so-called "repunit" prime.

A "repunit" (for "repeated unit") number is one that consists of nothing other than a sequence of 1's. The symbol R_n is used to denote the repunit number consisting of n repetitions of 1. So R_1 is just 1, R_2 is 11, R_3 is 111, and so on. Most repunit numbers are not prime, but a very small number are.

The first repunit prime is R_2, i.e., 11. The next is R_{19}, followed by R_{23} and then R_{317}. This last one was the largest known repunit number known when Oltikar and Wayland did their work. To obtain their 362-digit Smith number, you start with R_{317}, multiply it by 2, and then add on 45 zeros at the end. (It is easy to check that this is indeed a Smith number.) Had they but known it, however, they could have used the same repunit prime to break their own record. For you get another, larger Smith number if you take R_{317}, multiply it by 27, and then add on 361 zeros at the end. But this was not noticed until later.

What was realized at the time was that it would be possible to pull the same trick again with a larger repunit prime, if only one were known. Unfortunately, out of all the repunit numbers examined over the years, none were found that were primes. Until last year, that is, when it was discovered that R_{1031} is prime. [See Chapter 62.] (This is the only other repunit prime among the first 10,000 repunit numbers, incidentally.) From this prime number it is possible to construct a 1,178-digit Smith number. Double R_{1031} and append 147 zeros. For a bigger one still, multiply R_{1031} by 27 and append 1177 zeros. Or you can combine the two repunit primes R_{317} and R_{1031} to get a Smith number with 2,480 digits. Multiply together 162, R_{317}, R_{1031}, and then append 1131 zeros. (The digit sum in this case is 9279.)

But this is just small fry. When you really work at it there is much larger game to be had. Samuel Yates of Delray Beach, Florida, recently found a Smith number with a staggering 2,592,699 digits. And there you have the biggest known Smith number in the world. Or perhaps some *Guardian* reader can do better.

86

One is the number

March 19, 1987

Those readers old enough to have a well-thumbed book of logarithm tables might like to check the following rather curious phenomenon. The pages near to the front will show signs of far greater use than those at the back. Why should this be? Surely, given a range of problems requiring the use of logarithms, spread over many years, all pages should have had equal wear, should they not? But the evidence will undoubtedly point the other way. Numbers whose leading significant digit (lsd) is 1 seem to crop up far more than any others.

And it is not just logarithm tables that illustrate this phenomenon. Any book of physical constants or mathematical tables will show the same feature. And in the computer age, a daisy-wheel printer heavily-used for numerical work will wear out first of all on the digit 1. But how can this be? If you were to write down a (whole) number at random, any one of the digits 1 to 9 is equally likely to come first. So what is going on with numbers in tables? Surely they ought to have a random spread.

The simple fact is they do not. This was first pointed out to me several years ago by my colleague Dr. Peter Turner, at Lancaster University's mathematics department. Dr. Turner is one of a number of mathematicians who have investigated what is going on.

That there is a bias in the lsd of tabulated numbers seems to have been first mentioned in print by the American astronomer Simon Newcomb in 1881 in the *American Journal of Mathematics,* Vol. 4.

In 1938, Frank Benford wrote a report in the *Proceedings of the American Philosophical Society,* Vol. 78, in which he investigated the results of some 20,229 observations of various tables (lengths of rivers was one such, for example), and observed that the leading significant digits seemed to obey the following distribution: 1 in 30% of all cases, 2 in 17.6%, 3 in 12.5%, 4 in 9.7%, 5 in 7.9%, 6 in 6.7%, 7 in 5.8%, 8 in 5.1%, and 9 in 4.6%.

So almost one in three of all numbers in tables starts with a 1, while fewer than one in 20 has an lsd of 9.

See for yourself. For small tables the phenomenon may not be so marked. For example, see what you get with the multiplication table for all numbers from 1 to 9 by all numbers in that range (a total of 81 tabulated values). Then, using a computer, try a bigger multiplication table.

For more convincing evidence, where some really astronomical numbers are involved, of the values of $N!$ for N from 1 to 100, exactly 30 have an lsd of 1. Likewise for the values of 2^N for N in this range, or the values of the Fibonacci numbers U^N. [See Chapters 38 and 57.] Except for some obvious, trivial exceptions (such as addition tables) which may well have nothing to do with leading digits at all, any tables you draw up will show the same behavior.

The point is that mathematical operations—be they performed by man or machine or nature itself—do in fact produce a strong bias towards a low lsd, however much this may go against intuition. The lsd distribution is not spread equally over the digits 1 to 9, but follows instead of logarithmic pattern, with the proportion of numbers having lsd K given by the formula.

$$\log(K + 1) - \log(K).$$

(Logarithms are taken to base 10 here.) This formula gives rise to the percentage figures quoted earlier, with 1 occurring as lsd in 0.30 (i.e., 30%) of all cases, 2 in 0.176 (17.6%), etc. A small table like the multiplication table for numbers from 1 to 9 might not fit this formula terribly well, but other, bigger tables show an often remarkably close fit. Use a computer to investigate it for yourself, and you will see what I mean. Or do what Benford did and analyze some published physical tables. (For obvious reasons, a telephone directory will not work, as numbers are assigned in a specific manner.)

And if you want to see how your daisy-wheel printer is standing up to a long numerical job, take a look at the 1. That will get the most bashing.*

On a more serious note, the bias towards a lsd of 1 means that programmers have to be cautious in designing random number generators, where any bias must be eliminated. Usually, what is done in this case is to generate numbers much larger than the ones required, and take the random number from well inside those big numbers, dropping the first few digits. This solves the lsd problem, since the bias only occurs at the beginning of the number, not in the middle or end.

---

*Of course, the development of dot matrix and laserprinter technology has made this example now all but obsolete.

87

Infinite variety
April 16, 1987

Had *Horlicks** been available at the time, Hamlet, Prince of Denmark, might have been able to lay claim to taming the infinite. For as he says to Rosencrantz and Guildenstern in Act II, Scene 2, of Shakespeare's classic play, "O God, I could be bounded in a nutshell and count myself a king of infinite space, were it not that I have bad dreams." As it was, it is to the mathematicians of the nineteenth century that must go the large part of the credit for finally bringing the infinite under control.

The concept of infinity goes back a lot further than that, however. The ancient Greeks were certainly aware that mathematics must perforce deal with the infinite. Around 450 B.C., Anaxagoras wrote that "there is no smallest among the small and no largest among the large; but always something still smaller and something still larger." As soon as mathematics tries to deal with something as "commonplace" as the collection of all whole numbers, infinity has to be confronted. And at once a whole Pandora's Box of surprises opens up.

Hilbert's Hotel, for instance. This ultimate Hilton, were it ever to be built, would have to be in Texas, since it has an infinite number of rooms, each one numbered with a positive whole number, 1, 2, 3, and so on (ad infinitum). Such is the nature of infinity, that should you chance to arrive at this hotel on a night when every room is occupied, it would still be possible for the proprietor to give you a room without any guest being ejected. All that is required is for everyone already there to move into the next room (so the occupant of room 1 goes into room 2, room 2's occupant transfers to room 3, and so on), and at once room 1 becomes vacant for you to move into. Since there are infinitely many rooms, no one is in the position of having no room.

Or there is the paradox about the runner, formulated by the Greek philosopher Zeno some time in the fourth century B.C. This purports to show that a runner cannot cover a distance of (say) 100 meters, since before he can cover the whole he must first cover one half thereof. After that he must cover one half of what remains, then half of what is left, and so on. Always there remains a half of what is left to be covered. Since this process continues indefinitely, the runner can never cover the entire distance. Or so the argument goes.

In fact, paradoxes such as this are easily disposed of nowadays, now that we know how to handle the infinite. This particular problem has at its heart the infinite sum.

$$1/2 + 1/4 + 1/8 + 1/16 + \cdots$$

Though there are infinitely many terms in this

---

*A popular bedtime drink in Britain, purported to help one fall asleep.

sum, with each one equal to one-half the preceding term, it has a finite answer of 1 (representing the one hundred meters). What happens is that the manner in which the successive terms in the sum get smaller more than compensates for the fact that there are infinitely many of them to add together.

But things are not quite that simple. (Did I just say simple?) Consider the infinite sum

$$1/1 + 1/2 + 1/3 + 1/4 + 1/5 + \cdots$$

of the reciprocals of all positive whole numbers. This sum is known as the "harmonic series" because of a connection with musical scales. Although the terms in this sum grow smaller, they do so too slowly to compensate for the infinite number of them. As a certain Nicolae Oresme noticed in the 17th century, this infinite sum has an infinite answer.

And this is where the micro owner can have an evening's harmless fun. See how well your machine can cope with infinity by investigating how close it can get to infinity using the harmonic series. It is easy to write a short program to start adding the successive terms of this sum. Since the correct answer is infinity, your computer will not get that far. But just how far will it get?

Don't be too hopeful. This sum climbs up to infinity only very slowly. You need to add together 83 terms to get an answer greater than 5, and 12,367 terms to get beyond 10. A million terms gets you just past 14.3, while a billion brings you to 21 or thereabouts. To get the sum above 100, you need to add together

10,000,000,000,000,000,000,000,000,000,
　　000,000,000,000

(10 followed by 42 zeros) terms. Nevertheless you can have fun producing a table showing how many terms are required to reach various more modest subtotals.

You might also try adding together the reciprocals of all whole numbers that do not contain the digit 9. Curiously enough, though this sum would seem to involve "most" of the terms of the harmonic series, it does give a finite answer, somewhere between 22.4 and 23.3. (The exact result is not known.) On the other hand, the sum of the reciprocals of all the primes has an infinite answer.

And to finish, set your micro to work evaluating the infinite sum

$$1/1^2 + 1/2^2 + 1/3^2 + 1/4^2 + \cdots .$$

As Euler observed in 1736, the answer to this one is $\pi^2/6$. Once again, that amazingly versatile constant π enters the picture!

88

Sum election balance

May 14, 1987

They say you can prove anything with figures. Readers will no doubt have their own reaction to the sum displayed below. It is a particularly topical example of a so-called "cryptarithmetic puzzle." The challenge is to assign a single digit to each of the letters that appear in the sum (with every instance of the D getting the same digit, etc.) so that the resulting arithmetic sum is correct. There is just one way of doing this using digits other than zero. It is up to you to find them.*

```
        NEIL
     +  DAVID
     +  DAVID
     ───────
       MAGGIE
```

This problem was sent in by Keith Nelson of the Computer Sciences Company Ltd. of Slough in Berkshire, so supporters of tactical voting should direct their complaints to him, not me. [The answer is given in Chapter 89.]

On the same topic, the mathematics of three-way elections throws up some surprising results. For instance, suppose that one-third of electors rank the three parties in the order Conservative, Labour, Alliance, another third as Labour, Alliance, Conservative, and the remaining third as Alliance, Conservative, Labour. Then a solid two-thirds majority clearly prefer the Conservatives over Labour, and an equally impressive two-thirds majority prefers Labour over Alliance. On the face of it, this looks like bad news for the two Davids, apparently at the bottom of the pile, and good news for Maggie sitting at the top.

But no, if you check the figures you will see that this is not the case (at least in this example). An equally thumping two-thirds of the electorate prefer Alliance to Conservative!

What makes this curiosity of particular interest, of course, is that the scenario seems to be fairly close to what is actually going on, from which you may draw your own conclusions.

Continuing the theme of how unreliable intuition can turn out to be when numbers are involved, let's consider the second most-burning issue of the day: Aids. For the sake of argument, assume that the incidence of Aids in the population is 1 in 100,000. Suppose you were to take a test that has a 95 per cent chance of detecting Aids if it is there, and a 5 per cent chance of giving a positive result when it is not there. The test is positive, i.e., it says you have Aids. What are your chances of having the illness?

---

* This column was written during the run-up to a General Election in Great Britain. The leaders of the three main political parties in Britain at the time were Neil Kinnock (Labour Party), David Steele and David Owens (joint leaders of the Liberal-Social Democratic Party Alliance), and Margaret (Maggie) Thatcher (Conservative Party).

Most people think that the answer is 95%. In fact the correct answer—on the basis of the figures given—is that the positive test result would only imply a likelihood of having Aids of 0.019%, i.e., less than 1 in 5,000. Far less cause for alarm.

The confusion in cases such as this comes from ignoring what is a crucially important part of the data, namely the actual incidence of the illness in the population. If this is as small as, say, 1 in 100,000 as in the example here, its effect on the test result is highly significant. To obtain the correct result, what you should do is multiply this incidence rate by the likelihood that the test gives a positive result, in this case 95/5, which is 19.

Here is another example, investigated by the psychologists Kehneman and Tversky in the early 1970s. A group of subjects were told that an individual had been chosen from a group consisting of 70 engineers and 30 lawyers. The chosen individual had the following character sketch: "Dick is a 30-year-old man, married, with no children. A man of high ability and high motivation, he is likely to be quite successful in his field. He is well liked by his colleagues."

Given this evidence, would you say that Dick is likely to be an engineer, a lawyer, or equally likely to be either?

If you are typical of the test group, you will say that Dick is equally likely to be an engineer or a lawyer. But you would be wrong. Remember, Dick was originally chosen from a group consisting of 70 engineers and 30 lawyers, so the fact that the character sketch provides no deciding information means that the chances are 7 to 3 that Dick is an engineer.

Figures often mislead people. There is no shame in that: words can mislead as well. The problem with numbers is our tendency to treat them with some degree of awe, as if they are somehow more reliable than words (especially numbers that come out of a computer). This belief is wholly misplaced. Even honestly presented, undoctored figures and statistics can be hopelessly misinterpreted, let alone those that have been tampered with. The switching on of the election pundits' computers call for even more alertness than usual.

89

The thought machine

June 4, 1987

We all jump to conclusions. Occasionally the result is so dramatic that we (and often others) are made aware of that fact, possibly to our eventual embarrassment. But most of the time we correct any mistakes so quickly that we may well not notice the jump. Take the following simple little story.

"John was on his way to school last week. He was really worried about the math lesson. Last week he was unable to control the class."

Were you aware of the slight hiatus in your comprehension of this little tale when you got to the last sentence? For, as laboratory tests have shown, there will have been a slight but detectable delay as your mind quickly switched from its assumption that John was a schoolboy, to the realization that he was in fact the teacher. (Or was he? Perhaps the story proceeds somewhat differently.)

There, in a nutshell, you have one of the reasons why it is horrendously difficult to produce computers (and programs) that can understand ordinary speech. For humans, just a few words are enough to spark off a whole range of associations that aid understanding. Examples like the one I have just given demonstrate this fact by fooling the mechanism that generally works to our advantage— though even then it only takes a fraction of a second to make the necessary correction.

Or how about another example. "Max drove to the coast. The car ran well." Any-

thing strike you as odd? Probably not. Indeed, to a human being there is no difficulty in understanding these two sentences. But to a computer trying to grapple with it, what on earth is that car mentioned in the second sentence? Making sense of this involves a whole range of prior knowledge about what is involved in driving to the coast. As soon as we read the first sentence we know that there must be some vehicle involved—presumably Max's car—so the second sentence comes as no surprise. But for a computer this kind of thing can cause real problems: problems that are still a long way from being solved.

It is in order to come to grips with goals like the machine understanding of ordinary language that computer scientists are increasingly finding themselves turning to the work of psychologists, linguists, and philosophers. By examining the way people handle language, perhaps it will be possible to design computers to do the same thing.

In the meantime such work is starting to have a quite unexpected side-effect that some may find not a little unsettling.

It has turned out to be relatively easy to build systems that can perform— on equal or even better terms—the routine duties of an expert such as a doctor or the game playing of a minor chess master, but with the so-called simple tasks like boiling an egg or understanding everyday sentences the greatest difficulties arise. (Makes you think twice about experts, doesn't it?)

It also means that you should be suspicious of extravagant claims of some computer utopia lying just around the corner. It simply isn't that close. Mimicking the expert, who, by definition, is an expert in a small and highly restrictive area—looks impressive because of the status such people have in society. The truth is, however, that tieing a shoelace or understanding simple nursery rhymes are, from a computational viewpoint, far more sophisticated accomplishments than say, solving a problem in algebra.

Indeed, by far the most significant result that looks likely to come out of the computer revolution is a better appreciation of just what an incredible job nature has done in getting Homo sapiens to the state she has.

What got me thinking along these unusually (for this column) reflective lines was that I finally managed to get to the Theater Royal to see Hugh Whitemore's brilliant play *Breaking the Code,* in which Derek Jacobi's performance as the wartime code-breaker Alan Turing can only be described as stunning.

In one speech (taken, I believe, from a transcript of a talk given by Turing on the radio shortly before he died in 1954), the great computing pioneer speculates about computers built before the year 2000 being capable of thinking like (or better than) people. Well, it seems this particular prediction was hopelessly over-optimistic.

Finally, here is the answer to the cryptarithmetic puzzle that appeared in my last column [Chapter 88]. The problem was to assign nonzero digits to each of the letters A, D, E, G, I, L, M, N, V so that the sum

$$
\begin{array}{r}
\text{NEIL} \\
+ \text{ DAVID} \\
+ \text{ DAVID} \\
\hline
\text{MAGGIE}
\end{array}
$$

is correct. The only solution possible is:

$$
\begin{array}{r}
5246 \\
+\ 87348 \\
+\ 87348 \\
\hline
179942
\end{array}
$$

90

On and on into infinity

June 18, 1987

Call a positive whole number "sorted" if its digits are nondecreasing from left to right. So the number 1233489 is sorted but 1334288 is not, the problem in the last case being that drop down from 4 to 2. The following two questions about sorted numbers were posed by the Computer Science Problem Seminar at Stanford University in California.

First of all, let N be any number which consists of a string of 3s followed by a string of 6s followed by a single 7. Prove that N^2 is sorted. For example,

$$33\ 366\ 667^2 = 1\ 113\ 334\ 466\ 688\ 889.$$

When you have managed that, describe all numbers N for which both N and N^2 are sorted. (The kind of description you should look for is in terms of strings of digits much like the description in the first problem.)

And when you have done that, see what else you can discover about sorted numbers. You should not need any advanced mathematics for any of this. After all, it simply involves elementary arithmetic. But don't expect it to be easy, especially the part where you have to direct your own line of research. The reward in this case is all the greater, of course, and gives an insight into why people like myself devote our entire professional lives to pure research.

Which brings me to Geoff Beard from Merseyside. Following on from my column of

April 16 [Chapter 87] about the computation of partial sums of the infinite series $1/1+1/2+1/3+1/4+1/5+\cdots$ (which has an infinite answer if all terms are considered), Mr. Beard computed a table of values of the number of terms $K(n)$ required in order to obtain a sum greater than n. Thus, for $n=1$, $K(n)=1$; for $n=2$, $K(n)=4$; and for $n=3$, $K(n)=11$. These values are easily checked by hand. Using a computer or a good calculator, further values can easily be worked out. For instance, $K(4)=31$, $K(5)=83$, and $K(10)=12367$. Mr. Beard went up to $K(14)=675216$. After playing around with his table for a while he noticed that the ratio $K(n+1)/K(n)$ seemed to be getting closer and closer to the mathematical constant e, whose decimal expansion begins 2.71828. Was this just a coincidence, he asks?

The answer is no. Mr. Beard has stumbled on a mathematical fact that I did not mention in my original article. Readers with some knowledge of calculus might like to check for themselves why the ratio $K(n+1)/K(n)$ approximates the constant e as n increases. The clue is to regard the terms of the original infinite series as providing an approximation to the area under the curve $y=1/x$.

Finally, the following teaser, a variant of an old classic, was devised by Dr. David Singmaster of the South Bank Poly. An explorer leaves his base, heading due North, and walks for

10 miles, whereupon he plants a flag. Then he again heads North and walks a further 10 miles, which brings him back to his base. Assuming he walked in a straight line each time, whereabouts is his base on the Earth's surface? [Answer given in Chapter 91.]

91

Fermat's number is up

July 2, 1987

Fermat's Last Theorem is probably true—and that is official. What makes it official is that for the first time in 300 years there is some concrete evidence to support such a claim. Hitherto, the claim would have been based on nothing more than wishful thinking. Now all that has changed.

The Last Theorem, as it has come to be called, is probably the most famous unsolved problem of mathematics. Scribbled in the margin of a textbook by the 17th century French amateur mathematician Pierre de Fermat, this asserts that for any whole number n greater than 2, the equation

$$x^n + y^n = z^n$$

has no whole number solutions for x, y, z.

According to his marginal note, Fermat had discovered a "truly marvelous proof" of this, which, unfortunately, "the margin was too small to contain." (For the case $n = 2$ there are, of course, many solutions, the best known being $x = 3$, $y = 4$, $z = 5$. In this case the equation simply expresses a special case of Pythagoras' Theorem.) In all probability Fermat was mistaken in his claim—though he was, it must be admitted, one of the greatest mathematicians of all time. But it is still frustrating that no one since has been able either to prove or to refute the statement.

The first question facing the would-be solver of the Fermat problem is which way

to tackle the blessed question: true or false? Most people seem to favor the positive solution. But what evidence is there to support this? Some pretty impressive-looking computer results for a start.

Using a combination of sophisticated mathematics and brute force computing power, mathematicians have shown the Last Theorem is true for all values of n up to 125,000. And in 1983, a West German called Gerd Faltings showed that for any value of n there can be at most a finite number of (essentially different) solutions for x, y, z. [See Chapter 7.] But neither of these results amounts to anything like the kind of convincing evidence that is required in mathematics, which means that no one really knows whether or not the Last Theorem is likely to be true or false. And that was the situation until quite recently.

The difficulty with trying to decide whether the Fermat Theorem is true or not is that there are no underlying intuitions to fall back on. It is just a curious statement about numbers. (Indeed, the Faltings result mentioned a moment ago was obtained as a special case of some work in another area of mathematics). But now, thanks to some work by Gerhard Frey of the University of the Saarlands in Germany, a connection has been established between the Last Theorem and another branch of mathematics that is sufficiently well established to enable intelligent conjectures (rather than wild guesses or wishful thinking) to be made.

The idea is to start by assuming that there is a value of the exponent n for which the Fermat equation does have a solution, in which case there will be a prime value of n, say $n = p$, with a solution. Call this solution $x = a$, $y = b$, $z = c$. So a, b, c, and p are assumed to satisfy the identity

$$a^p + b^p = c^p$$

Frey's key step now is to look at the equation

$$y^2 = x(x - a^p)(x - c^p)$$

This is a special case of what is known as an "elliptic curve," which belongs to an area of mathematics about which there is a wealth of information. So much, in fact, that experts in the area were soon convinced that an elliptic curve with the properties Frey's had simply could not exist.

This would at once prove Fermat's Last Theorem, of course, since Frey obtained his equation by assuming that the Fermat equation did have a solution.

What remains to be done is to discover a rigorous proof that such an elliptic curve is impossible. So far no such proof has emerged, but at least mathematicians are now able to work in an area of mathematics where there are many powerful tools available and where it is possible to rely upon mathematical intuitions to support one's approach. Suddenly,

Fermat's Last Theorem looks a step closer to its final resolution.

And so to that tantalizing little problem in my last column (June 18 [Chapter 90]). This concerned the explorer who leaves his base, heading due North, and walks for 10 miles. Whereupon he plants a flag and then again heads North and walks a further 10 miles, bringing him back to his base. The problem was to say where his base is. (The question stated that he walks in a straight line each time.)

The key is to pay close attention to the way the question is worded. On each of his two walks he starts off heading North, but there is nothing to say he continues to do so. So if his base is anywhere within 10 miles of the North Pole, two straight-line 10-mile walks that each start out towards the Pole will bring him back to his base, crossing the Pole twice in the process.

Update. Soon after Frey's result, Kenneth Ribet of the University of California at Berkeley showed that the possibility of a connection between elliptic curves and Fermat's Last Theorem raised by Frey's work was a genuine connection.

In June 1993, Dr. Andrew Wiles of Princeton University announced that he had solved Fermat's Last Theorem in precisely the manner outlined in this article. See Chapter 143 for details.

92

A clever little number

July 16, 1987

Sitting in a family living room in a house in Paramus, New Jersey, between the couch and the armchair, is a cobbled-together piece of computing equipment that already holds a number of world records, and that, for the kinds of computation it was designed for, is only about 10 times slower than the world-beating Cray-1. This makes it pretty well equal to anything that comes out of IBM, and some 30 times faster than a late 1970s mainframe computer. But cobbled together? That is unkind, for the real secret of this amazing piece of equipment is its skillful design.

The story begins back in 1980. Mathematician Harvey Dubner and his design engineer son Robert ran a small outfit in New Jersey called Dubner Computer Systems, Inc., which specialized in the design and construction of custom computer systems. Using some of the circuit boards they had developed for contract jobs, the two Dubners put together a fairly powerful micro using a 2MHz 8080 processor, 48K of RAM, and 10K of ROM. In assembly language they wrote some special routines for attacking number theory problems involving numbers with up to 200 digits. It worked well, but by no means well enough to get them into the record books. In particular, one of their first main goals was well out of reach.

This was to verify that the repunit number R_{1031} was prime. (The number R_n consists of n consecutive 1s. For values of n less than 10,000, the only such numbers that are prime occur for n equal to 2, 19, 23, 317, and 1031. In 1980, all that was known about R_{1031} was that it was likely to be a prime). The size of this number put it well outside the range of the Dubner's homemade machine, so they started to improve it.

Changing the software alone increased the computing speed by a factor of seven. Then, in 1981, they added a pair of Multiply-Divide-Units, an inexpensive part made by RCA. This speeded up computations by a further factor of more than ten, and brought 1,000-digit numbers within reach—just.

But already even greater improvements were on the way, in the form of the specially-built CBG-2 Character/Background Generator (essentially an 8080 processor with a grafted on 20-bit wide bit-slice microprocessor) and a TRW MPY 16 (a 64-pin parallel multiplier that, for an initial outlay of a few hundred dollars allows the multiplication of two 16-bit numbers in less than 150 billionths of a second). In its new form, their sitting room computer was operating some 5,900 times faster than the original version.

Now the Dubners were in business. Handling numbers with one or two thousand digits became child's play. For instance, squaring a 1,000-digit number took a mere 1.6 thousandths of a second, which was about 2,700

213

times faster than an IBM PC. And before long it put the Dubners into the record books.

In collaboration with Hugh Williams of the University of Manitoba, Harvey Dubner used his machine to obtain the long sought after proof that R_{1031} was prime [Chapter 62], then to show that there were no other repunit primes R_n for n less than 10,000. And more . . .

The Dubners found

- the largest known palindromic prime number. It starts and ends with a 1, has a single 5 in the middle, and zeros everywhere else for a total of 2,977 digits;
- the biggest known prime of the form $N! + 1$ (primes of this form have been of interest since 350 B.C., when Euclid used such numbers to prove that there is an infinitude of prime numbers). It is the 4,042-digit number $1477! + 1$
- the biggest known "prime-factorial plus one" prime number, the 2,038-digit giant

$$2 \cdot 3 \cdot 5 \cdot 7 \cdot 11 \cdots 4787 + 1.$$

(These numbers are of interest to those whose textbook gives Euclid's proof in another fashion!)

- the largest known twin-prime pair (two primes separated by 2): they are the two numbers adjacent to the even number that starts off 107570463 and then continues with 2,250 zeros;
- the largest prime number not of the special Mersenne form $(2^n - 1)$ used to obtain the overall world records. Dubner's "ordinary" record prime has 8,500 digits.

Admittedly the computations take some time—around 40 days to find the record "ordinary" prime, for instance. But when you think that all that is involved is a home computer equivalent with five or six hundred dollars worth of specialized add-on parts, with the whole apparatus sitting quietly in the Dubners' living room, you begin to realize that there is a moral here.

When it comes to performing highly specialized computations (the Dubners' machine is only suitable for whole number arithmetic), then dollar for dollar, a tailor-made computer can easily outstrip the biggest mainframe. Buyers of expensive supercomputers should perhaps ask themselves whether they are paying for a host of features they do not need. It might be better to hire a good computer engineer instead. For the cost of a box of specialized chips, you could save $12 million or so.

93

Putting your foot in it

August 27, 1987

At about this time of year, *Micromaths* usually puts aside its normal fare in favor of something more in keeping with the holiday period. So here goes. As usual there are no prizes for getting the questions right.

1. Find the only four whole numbers equal to the sum of the cubes of their digits.

2. Find the only pair of digits such that putting a decimal point between them produces their average.

3. It is an indisputable fact that if you were to go around your neighborhood with this quiz and record each person's resulting performance along with their shoe size, you would find an unmistakable statistical correlation between quiz score and foot size. Explain this phenomenon.

4. Find digits A, B, C such that the 3-digit number ABC is equal to the sum $A! + B! + C!$ (where $A!$ is the product of all the numbers from 1 to A inclusive.).

5. Given six glasses $A, B, \ldots, F$ in a row (in that order) with A, B, C full and D, E, F empty, your task is to arrange for alternate glasses to be full by moving just one glass. How do you do this?

6. Arrange the digits 1, 2, 3, 4, 5, 6, 7 into an addition sum with answer 100.

7. Rearrange the letters of NEW DOOR to form one word.

8. Why are 1986 pennies worth almost £20?

9. It is known that the center of a regular tetrahedron and any two vertices all lie in the same plane. Is this true for an irregular tetrahedron?

10. After a bridge hand has been dealt, which is the most likely, that you and your partner hold all the clubs or that between you you hold no clubs?

11. "Smith has over a hundred books," says Jones. "No, he has fewer than that," Brown replies. "Surely he owns at least one book," comments White. If only one of these statements is true, how many books does Smith have?

To prevent you inadvertently glancing at the answers to the above, here are a couple of mathematical oddities to keep the two apart.

Remember being perplexed by that "imaginary" number i, the square root of -1? Well, how about this for an unexpected result? If you take the imaginary number i and raise it to the power i (in other words, calculate i^i), then the answer is a real number, equal approximately to 0.20788.

Spell out the name of any number (whole, fractional, real, or complex) in the normal fashion, and then, starting with your chosen number, iterate the procedure of counting all the letters in the name and writing out this number in English. You will eventually end up with the repeating sequence FOUR, FOUR, FOUR, ...

Answers

The only quiz questions whose answer will be given here are numbers 3, 5, 7, 8, 9, and 10, since each of these involves some kind of a twist.

3. If you give this test to everyone in your neighborhood, there would be a fair number of babies and very small children, whose small feet and zero quiz score would impose a spurious relationship on your data. Which reminds me of the market researcher who approached a well-known statistician carrying a huge sheet of graph paper on which he had painstakingly plotted a large number of points.

"I was wondering whether it is possible to put a straight line through this data," mused the market researcher. The statistician stared at the document thoughtfully for some time before replying solemnly:

"If you ask me, that would be by far the best thing to do with it."

5. Pick up glass B, pour its contents into glass E, and then return glass B to its original position.

7. The rearrangement you are seeking is ONE WORD.

8. 1986 pennies add up to exactly £19.86p, which is just 14p short of £20.

9. Yes. Any three points will lie in the same plane.

10. The two events are equally likely. If you and your partner hold all the clubs, your two opponents hold none, and vice versa, so the situations are entirely symmetrical.

94

Down the tubes

September 24, 1987

Give it a complicated mathematical calculation to do, and a modern computer will perform wonderfully—far better than a human. But try to get it to do something that people do with ease, like recognize Snoopy as a drawing of a dog, and you may have to wait for weeks for the result. For it is with the seemingly trivial everyday tasks such as recognition of simple scenes or understanding ordinary language that existing computer technology comes up against its limits.

Surmounting these limits, if this is at all possible, will undoubtedly involve not only new programming techniques, but new kinds of computers as well. Computers that operate in a manner closer to that of the human brain than do existing machines. For which enterprise it is reasonable to start off by taking a look at just what is the difference between the human brain and a modern electronic computer. Just what is it that makes the human mind so powerful at performing certain tasks that leave present-day computers reeling?

The basic computational element out of which a present-day computer is constructed is the transistor, nowadays etched into a wafer of silicon. The brain equivalent is the neuron. Does the power of the brain lie in the fact that the neuron operates much faster than a transistor? The answer is an emphatic "No."

The transistor can switch (i.e., alter its internal state) about a million times faster than the millisecond or so it takes for a neuron to switch. Nor can it be that the brain contains significantly more basic devices. The 10^{10} or so neurons in the brain compares well enough with the 10^9 or thereabouts transistors in a modern mainframe, but when the switching times are taken into account you see that while the computer is capable of some 10^{18} switching events per second, the human brain can only manage about 10^{13}. In other words, in terms of raw, basic computing power, a typical modern mainframe would seem to be some 10,000 times faster than the brain.

This figure is, of course, nonsense. When it comes to performing complicated mathematical calculations, the computer outperforms the human brain by a factor way in excess of this, while for such tasks as picture recognition the human brain easily outstrips the fastest mainframe. So what makes all the difference? The answer is that the number and speed of the individual computing elements is only part of the story. How they are connected together is at least as (and probably much more) important. A number of factors, among them the engineering constraints that governed the design of the very first computers in the 1940s, have resulted in a computer industry that has for the greater part followed practically to its limits one particular, very narrow path: the so-called von Neumann architecture.

When the early computing pioneers such as John von Neumann (in the US) and Alan Turing (in the United Kingdom) were designing the first electronic computers, they were constrained by a fundamental cost differential. The fast devices (such as vacuum tubes) used to construct the central processing units—what we now know as the CPU, the part where the actual computing is done—were very expensive, whereas the delay lines or storage tubes used for data storage—what we now call memory—were much cheaper. So it made sense to design computers in a way that kept those expensive vacuum tubes as busy as possible. Since this division of activity into processing and memory fit well into some of the theoretical work on computing that had preceded the engineering practice, it is not surprising that it led to a whole generation of computing personnel thinking that this was the way computers had to be.

But consider the situation today. No longer do we require different materials in order to construct our processors and memory units. Everything is made from silicon. A typical modern mainframe contains about a square meter of the stuff. Of this, the CPU may occupy some two or three percent. This means that the classical von Neumann computer architecture results in some 97% of the machine effectively sitting idle at any one time. And at a cost of around a million dollars a square meter (in its processed form) this is not exactly a sensible use of resources.

The bigger we make our machines, the worse the problem becomes, as increasing amounts of time are spent shifting data around inside the computer in order to keep the all important central processor happy. What began as a design expediency has now become a problem.

It even has a name. It is called the von Neumann bottleneck. The solution is, in principle, obvious. If both processing and memory take place in essentially the same physical environment, design your computer so that processing takes place throughout the entire machine—in the "memory" itself, to use present day terminology. In other words, abandon the von Neumann design and take a step closer to the human brain.

Fanciful dreaming? Not at all. The first step has already been taken, with the construction of the world's first "Thinking Machine." In my next column [Chapter 95], I will describe how it was done.

95

Making the right connections
October 8, 1987

Present-day computer technology is coming very close to its limits. In order for there to be any significant progress on any of a number of problems, new kinds of computers will have to be designed and built. It may even be necessary for us to completely rethink the way computers fit into our society. So, among others, says W. Daniel Hillis, founder of the US-based Thinking Machines Corporation.

As explained in my last column [Chapter 94], the major obstacle to any fundamental increase in computing power lies in the so-called "von Neumann bottleneck": the basic computer design that effectively separates data processing (the CPU, or Central Processing Unit) from data storage (the memory).

The operating cycle of a typical present day computer consists of selecting items of data from memory, transferring them to the CPU, operating on them (e.g., adding together two numbers or comparing two data items to see if they are the same), and then possibly returning the result to some further storage location in memory. It is their ability to perform this simple cycle so very fast (say tens of thousands of times in a second) that makes electronic computers as useful as they are.

Broadly speaking, there are two reasons for adopting this design of computer. One is theoretical, in that it provides an easily understood mode of operation that makes programming a reasonably straight-forward mat-ter. The other reason lies in the engineering and financial constraints that prevailed when the very first computers were built in the 1940s. The fast switching devices such as vacuum tubes that were required for the construction of processing units were very expensive, whereas the delay lines and cathode ray tubes used for memory were comparatively cheap. So it made financial sense to have a single processing unit served by a large memory bank.

Nowadays, however, both memory and processor are made from the same stuff: silicon. There's about a square meter of it in a modern mainframe. In this kind of world, the von Neumann processor/memory division means that at any one time something like 97% of that highly expensive silicon circuitry (the part designated as "memory") is sitting idle, while only some 2 or 3% (the "processor" part) is working.

At around a million dollars per square meter, this is hardly a profitable arrangement. Besides, it results in a computer which, though well suited for straightforward numerical work, is hopelessly ill-equipped for such seemingly simple everyday tasks as recognition of a visual scene or understanding ordinary language.

As numerous studies have confirmed, what is required for problems of this nature is not a single, powerful processor working very fast, but a large number (maybe tens of thousands, or more) of fairly modest basic computing units all working together at the same time in parallel.

The first difficulty encountered in trying to design a "parallel computer" is how to link together the individual processors involved, for it is only by the individual processors being able to communicate effectively with each other that the entire structure may achieve its computational power.

Various possible designs immediately suggest themselves. The simplest would be to connect together all the processors in a straight line. But then there will be a long delay whenever two distant processors wish to exchange information. Linking together (say) 64 processors in this way would require any end-to-end message to pass through each one of the 62 intervening ones, resulting in a performance worse than the old von Neumann design!

Linking the processors together in the form of a ring would improve matters by a factor of 2, of course, since now the maximum distance any message would have to travel would be through 31 intervening processors. Still better, how about a rectangular 8×8 grid? Here the maximum message distance is only 16 steps long (from corner to opposite corner). Faster again, a $4 \times 4 \times 4$ cube, where the maximum separation between processors is a mere 12 units.

But why stop there? Though it is not physically possible to construct, say, a four-dimensional "hypercube," such an object is mathematically perfectly feasible, and there is nothing to prevent the interconnection of an array of computing devices as if they were arranged in such a hypercube. (The essential point here is that the "wires" can bend around.) And the higher the dimension, the better the resulting network becomes.

This is the approach adopted by W. Daniel Hillis in the design of what he has called the Connection Machine. Taking full advantage of the essential sameness of all modern computing elements, processor and memory, he took 65,536 ($= 2^{16}$) basic processing units, each of which has control over its own, "tiny," personal 4096 ($= 2^{12}$)-bit memory, and linked them together in the form of a 12-dimensional hypercube. With this architecture, no two individual processors are more than 12 steps apart, so communication between them is very fast indeed. But now there are over 65,000 of the blighters, all able to compute on their own (using their little personal memories), and all interlinked.

Physically, the entire apparatus consists of 128 circuit boards arranged to form a (real, three-dimensional) cube measuring one and a half meters each way. Each board contains 32 fairly ordinary silicon chips, on each one of which are etched 16 of the basic processing-memory units and a specially developed routing device that controls the way the individual processors communicate.

And it works!

After developing the Connection Machine as part of his PhD work at the Massachusetts Institute of Technology, Hillis founded Thinking Machines Corporation to build and sell his creation. So far around 20 of them are in use. Designed to be used in conjunction with a conventional computer as a "host," the Connection Machine is already promising to open up entire new vistas in computing, with advances in image recognition and text analysis being the likely first benefits. A specially developed form of the "artificial intelligence" programming language LISP enables programs to be written.

96

Silicon Valley scholars

October 22, 1987

California's Silicon Valley must surely contain the biggest concentration of mathematical and computing talent in the world. Stanford University, in the heart of the valley, is generally regarded as one of the top four or five universities in the US as far as computer science is concerned (the other obvious contenders being the University of California at Berkeley, just a few miles away across the San Francisco Bay, the Massachusetts Institute of Technology, Carnegie-Mellon University in Pennsylvania, and the University of Texas at Austin). On the industrial side there are the huge research and development laboratories of such firms as Xerox, SRI, Hewlett-Packard, Fairchild, Apple, DEC, and IBM. To say nothing of NASA's Ames Research Center, whose vast, cavernous hangar building, once home to giant airships, dominates the Bay shoreline just to the south of Stanford like some giant beached whale.

Since a great many of the amassed high tech brains that sustain this twentieth century empire have school age children, it would be reasonable to assume that the local schools too are somewhat above average where mathematics is concerned. Or so, at least, my wife and I told ourselves when we dragged our two offspring from their excellent British schools to accompany me on my year-long trip to Stanford, just begun.

We need not have worried. The standard of education at the ordinary state schools in Palo Alto (the city where Stanford is located) is every bit as good as any you could find anywhere in the world. When it comes to mathematics ... well, judge for yourself. The following is the mathematics homework sheet that my 13-year-old daughter brought home on her first day at her new school. See how well you do at it. (The more timorous reader can safely skip over this part and pick up at the Micromaths Sports Section that follows.)

1. The picture below shows a 5 by 9 grid made up of 45 small squares. How many squares are there altogether in this diagram?

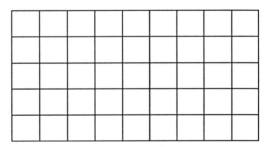

FIGURE 31. **The Silicon Valley Puzzle.** How many squares do you see?

2. How many digits are needed to number all the pages of a book having 725 pages?

3. What is the one's digit in 2^{999}?

4. If $A * B$ means $4A - 3B$, find the value of $(7 * 4) * 2$.

5. Find the next three numbers in the sequence $1, 4, 9, 16, 25, 36, 64, \ldots$.

6. How many different outfits can you make with 4 pairs of pants, 3 shirts, 3 pairs of shoes, and 2 hats?

7. How many car license plates can you make, each consisting of 3 letters?

8. How many consisting of 3 letters but not starting with an A?

9. How many consisting of 2 letters followed by 4 digits?

Continuing this week's All-American theme, the Science and Computing section of the *New York Times* on September 8 reflected on just what it is that makes America tick by kicking off (if that is the right expression to use in this context) with a fascinating report on some recent computer-aided research into baseball pitching.

Much as occurs in cricket, baseball enthusiasts are forever arguing the skills and merits of various legendary pitchers able to impart on the ball some miraculous force that causes it to change direction in flight several times so as to defeat even the most brilliant hitter. To the more skeptical, aware of some of the basic laws of physics, most of the claims made about pitching prowess tend to be taken with a large pinch of salt. But now, with the help of some wind tunnel experiments and a sophisticated computer simulation, researcher Joel W. Hollenberg of the Cooper Union School of Engineering in New York has come up with some concrete evidence to back up some of those old fans' stories.

Concentrating on the so-called "knuckleball," a difficult-to-execute but notoriously deceptive pitch, Hollenberg's computer program has demonstrated that the double change in direction the ball exhibits in its relatively sedate flight is caused by minute changes in the air flow around the ball's surface. The rough stitches of the seam create a complex pattern of turbulence that varies as the slow spin imparted to the ball by the pitcher gradually rotates the seam in flight.

The trick (or skill) in throwing a good knuckleball is to achieve just the right amount of spin. Too much and, like a gyroscope, the ball's flight will be straight and true. No spin at all and the ball's stitching will create a sideways drag that causes the ball to veer to one side but not to change direction a second time.

But give it just the right amount of slow spin and you can follow American baseball heroes into the Hall of Fame.

97

Prime chops

November 5, 1987

As regular readers of this column doubtless know, a prime number is a positive whole number that can only be exactly divided by itself and the number 1. Examples of prime number abound: 2, 3, 5, 7, 11, 23, and the number that starts with a 1, ends with a 7, and has 60 zeros in between.

There are infinitely many prime numbers (this was known to Euler as long ago as 350 B.C.) which means that there is unlikely to be any end to the game of hunting for world record largest known primes, a game that is nowadays played on the world's most powerful known computers.

Away from the limelight of the supercomputer boys, there are still many little quests involving prime numbers that can keep the home micro enthusiast busy as the evenings draw in. Though the world record prime number hunters use sophisticated mathematical methods in order to test numbers to see if they are prime; the problems presented here can all be handled using a very simple test.

Just write a routine that tests a number N to see if it is a prime by dividing N by each of the numbers 2 and all odd numbers up to the square root of N, checking in each case to see if the result is a whole number. (If it ever is, you know N is not a prime and can stop the search. If the square root of N is reached without an exact divisor being found, then N must be prime.)

For example, there are the so-called *permutable primes*. These are prime numbers that remain prime when you rearrange their digits in any order you please. For example, 13 is a permutable prime, since both 13 and 31 are prime. Again, 113 is a permutable prime, since it and each of the numbers 131 and 311 is prime. It is known that there are only seven such numbers within reasonable range (less than about 4 followed by 467 zeros, in fact). You now know two of them. Find the other five.

Or there are the *truncatable primes*. These are prime numbers that remain prime when you repeatedly chop off the initial digit.

For example, the permutable prime mentioned above, 113, is a truncatable prime, since both of the numbers 13 and 3 are primes. At least one example is known of a truncatable prime with 24 digits. See how well you can do. (Numbers that contain a zero are not allowed, by the way. Otherwise the giant prime 1 followed by 60 zeros followed by 7 would win hands down.)

Moving away from primes, but still staying with problems about numbers, we finish this week with a simple task related to the famous Fibonacci rabbits sequence. This, you may recall, concerns the numbers of pairs of rabbits you will have each month if you start off with one young pair and let them breed monogamously (though incestuously) from then on according to the rule that each pair takes one month to reach maturity and each adult pair produces one new pair each month. (Remarkably, no rabbits die. This can only happen in Mathematical Wonderland.)

It does not take long to figure out that the monthly populations (in pairs of rabbits) is given by the number sequence

$$1, 1, 2, 3, 5, 8, 13, 21, \ldots$$

where each number in the sequence is equal to the sum of the two numbers before it.

Leaving Fibonacci's rabbit problem now; the same idea can be used to create more general Fibonacci-type sequences. Starting off with any two pairs of numbers, you can generate a sequence by the same procedure of always adding together the final two numbers to give the next one.

For example, if you start off with the numbers 4 and 7, you get the sequence 4, 7, 11, 18, 29, 47, ...

And here a rather curious thing has occurred. Sitting there among the numbers in the sequence is the number 47, whose digits are exactly the two digits you started with (in exactly the same order). How many other examples can you find where this happens?

When you have sorted that one out, you can start on an endless process of generalization. Starting with a 3-digit number, write down its three digits and use them to produce a Fibonacci-style sequence where, instead of adding together the final two numbers to give the next, you add together the final three. For instance, starting with the number 123 you would get the sequence 1, 2, 3, 6, 11, 20, 37, ...

In this case, if you continue on, you will find that the number 123 itself does not appear in the sequence. There are only two 3-digit numbers that reappear in the 3-fold Fibonacci sequence which starts off with their three digits (in order). What are they?

Then maybe you would like to take a look at 4-digit numbers (along with 4-fold Fibonacci sequences). And so on. And if you get past 7-digit numbers, let me know. You will then be entering the record books. To date the largest number known that appears in its own digit-fold Fibonacci sequence is 7,913,837.

Good hunting!

98

Damned lies

November 19, 1987

This twenty-word sentence, the opening sentence in the *Micromaths* column in today's *Guardian,* contains a number of self-references.

For one thing, the sentence describes a feature of its own syntax, namely it tells you how many words it contains. It also takes note of the fact that it is the first sentence of the article, that it appears in the *Micromaths* column, and that the column appears in the *Guardian* for today. And then, on top of all that, it makes the point that it is a sentence which refers to itself in several different ways. (One of these ways being, of course, the fact that it is self-referential.)

The point of starting with these remarks is that on the last occasion when *Micromaths* ventured into the thorny territory of the ancient Liar Paradox [Chapter 83], it sparked off a flood of mail to the effect that self-reference was in itself something that civilized man ought not to contemplate. Or at least, something that was not a proper topic for mathematics.

Neither viewpoint is, of course, tenable. As today's opening sentence is meant to indicate, it is quite possible to make a perfectly understandable assertion that refers to itself. (Indeed, such occurrences are not particularly rare.) And since the "job" of mathematics is to study the underlying logic of everyday phenomena, it is just not acceptable to declare some area of everyday life "off limits" for mathematics. What the mathematician must do is figure out just what it is about self-reference that causes problems.

And there can be no doubt that there are problems, of which the classic Liar Paradox is perhaps the best known example. This is where someone stands up and says:

"What I am now saying is false."

The question is, is this a true claim or a false one? If it is true, then according to that true claim, it is false. And if it is false, then it must be the case that what is being said is true; that is, the claim is true. So, if it is true, it is false, and if it is false it is true. A genuine paradox.

Moreover, it is a paradox that will not be made to go away simply by declaring self-reference to be *verboten*. Since self-reference is not in itself a contradictory thing (witness the opening sentence), the very least that must be done in this direction is to say just what it is about self-reference that causes the problems in examples like the Liar Paradox.

Is it perhaps the combination of self-reference and truth? Not really. It is easy to modify today's opening sentence to include a reference to its own truth, and still have a perfectly informative, noncontradictory assertion.

Or perhaps the difficulty lies in our expectation that a sentence like the Liar must be either "true" or "false"? Maybe sentences like these have some sort of indeterminate "truth" status, neither true nor false (call it middling?). In which case, what happens when someone stands up and says "This sentence is either false or middling"? (I'll leave you to figure this one out. It seems to be just as paradoxical as the original Liar sentence.)

Such questions, once thought to be the province of the ivory tower philosopher, are nowadays discussed in university computer science departments and large computer research establishments. And you do not have to look very hard to see the reason.

Given the enormous complexity of today's computer systems, the only hope of keeping track of what is going on once the machine has been plugged in, is to perform a mathematical analysis of the system. "Semantics" is the name given to such a study. Loosely speaking, it means providing a mathematical explanation of what the system's activities "mean."

Providing the mathematical tools necessary for such a study is one of the tasks facing the so-called "theoretical computer scientist." It is a branch of mathematics that is still very much in its infancy. Trying to come to grips with phenomena such as the Liar Paradox is a marvelous test-bed for any new theories.

Indeed, one of the most recent attempts at resolving the Liar arose directly from some work in computer science carried out by Manchester University researcher Peter Aczel a couple of years ago. His work on self-referential computer programs was immediately seized upon by two Stanford University based philosophers, Jon Barwise and John Etchemendy, and developed to provide a new resolution to the paradox.

The Barwise–Etchemendy solution is described in their new 180-page book just published by Oxford University Press, entitled *The Liar.* It is a book well worth reading for anyone with an interest in puzzles such as the Liar Paradox.

In essence, what the book's authors discover as a result of the mathematical analysis they carried out using Aczel's ideas, is that the root of the paradox appears to lie in a difference between something being true and our having the information that it is true. One of the things that is proved in the book is that while the Liar sentence is false, it is just not possible to have the information that this is the case.

Of course, as expressed here, this does not really tell you very much. (After all, since I have just told you that the Liar sentence is false, surely you now do have the information that this is so.) But then Barwise and Etchemendy take 180 pages to supply the full picture, not just one sentence.

99

Computer dating challenge

December 3, 1987

Making a date with your computer takes on an entirely new meaning with the first of this week's problem round-up. It all starts with a computer teaser due to the California-based mathematician Charles W. Trigg. In a short article in the *Journal of Recreational Mathematics* earlier this year (Vol. 16, No. 1, pp. 10–11), Trigg observes that the squares of certain 4-digit numbers embed recent years. For example, if you square 4458 you get the answer 19873764, which starts off with the year 1987. This is by no means the only such example, and in his article Trigg lists several similar numbers. some of these start off with the year number, others just contain the year number somewhere within them, either in the middle or at the end. Writing a computer program to find other examples, say for the years 1980 to 1990, should present no insurmountable obstacles to regular readers of *Computer Guardian.*

More of a challenge is to find 4-digit numbers whose square gives a complete date of historical significance. The format for the date could be in the form either year-month-day or day-month-year. So today's date would be either 19871203 or else 03121987. Neither of these happens to be the square of a 4-digit number. But how many important dates can you find that are such squares?

The second problem requires nothing more than a calculator or a table of square roots. The task is to find numbers whose square roots may be expressed as an arithmetic sum of their digits, taken in order. For example, the square root of 81 is 9, which is equal to $8 + 1$; the square root of 361 is equal to $3 \times 6 + 1$; the square root of 2025 is equal to $20 \times 2 + 5$; and the square root of 8281 is equal to $82 + 8 + 1$.

A variation on the same theme is to allow the order of the digits to change. For instance, the square root of 169 is equal to $19 - 6$, where the three digits 1, 6, and 9 are the same but with a different ordering.

A similar problem is to look for what are known as Kaprekar numbers. These are numbers whose square is equal to the sum of the two numbers you get by splitting the number into two halves. (In the case of a number with an odd number of digits, there are two ways to split it, and either one is allowed.) So, for example, 45 is a Kaprekar number, since $45^2 = 2025$ and $20 + 25 = 45$. And 297 is another one, since its square is 88209 and 88 plus 209 gives 297 again.

One curious observation about Kaprekar numbers is that the "magician's" number 142857 is one. This is the number that you get in the form of an infinite repeating cycle if you work out the decimal expression for $1/7$:

$$1/7 = 0.142857\ 142857\ 142857\ldots\ .$$

Multiplying this result by any one of the numbers 2 to 6 (to give the decimal expression for each of 2/7 to 6/7) results in the very same sequence 142857 shifted along a bit. (Try it and see.) [See Chapter 12.]

Finally, A. Annett writes from Clevedon that an item in *Micromaths* on August 27 [Chapter 93] has left him with some sleepless nights. In that article I mentioned that if you take the imaginary number i (the square root of -1) and raise it to the power i, the result is the real number 0.20788 (to 5 places of decimals). "How is this answer arrived at?" Mr. Annett asks.

Well, like most things involving the imaginary quantity i, you have to make use of advanced mathematics. So hang on and away we go.

It all depends on the identity

$$e^{ix} = \cos x + i \sin x$$

where the $e = 2.71828\ldots$ is the base for the natural logarithms. If you take x to be the angle $\pi/2$ in this identity, then $\cos x = 0$ and $\sin x = 1$, so the expression on the right takes the value i. This gives the formula

$$i = e^{i\pi/2}$$

for i. So, raising this expression to the power i gives

$$i^i = (e^{i\pi/2})^i.$$

Since $i^2 = -1$, this simplifies to give $i^i = e^{-\pi/2}$. And now you can let your calculator take over.

So now you know. And if you are still with me, you can use the same formula to work out real answers to other exponentials involving the imaginary square root of -1. What, for instance, does $e^{i\pi}$ work out to be? Since e and π are the two most important constants of mathematics, you would expect the answer to be something very special indeed. And so, in fact, it turns out to be, as you will discover when you work it out.

100

Back to key one

December 17, 1987

When you sit in front of your computer and watch it produce the output from your latest programming effort, it is hard not to regard the machine as somehow doing what you told it to do (assuming, of course, that the output is what you expected to obtain). You can imagine the computer working its way through your instructions one by one, adding two numbers here, taking a square root there, making a decision somewhere else.

But the machine does not see any of this. All it does is route electric currents around in a complicated, though at heart very basic way. In physical terms, that is all that is going on. It starts the moment you switch on the computer, before you have written a single line of instructions, and it continues long after your program has run, right up to the moment when you hit the off switch.

So where exactly is your program, and just what is it? This question is one that few computer scientists ever worry about. They simply dispose of the problem by giving it a name: *software*. "Soft" is meant to indicate the "ethereal" nature of things such as programs (which come in a whole variety of levels from the most basic, binary microcode all the way up through high level programming languages to sophisticated packages that handle everyday language).

By contrast, the *hardware* is the physical aspect of computing, the part that can be ana-

lyzed in terms of electrons flowing along channels and so on, the part that is really there as a physical lump sitting before you.

But faced with seemingly insurmountable difficulties in trying to realize the sixties dream of artificial intelligence (even in its slimmed down eighties manifestation), a small but growing number of workers are beginning to take a serious look at just what it is about a computer program that manages to give significance to that myriad of flowing electrons inside the machine.

"Just what," they ask, "is required in order for a particular internal configuration of the computer (i.e., a particular setting of the various gates that make up the machine) to represent some feature of the world (say the positions, speeds and directions of aircraft in the sky)? And what is the link between a series of changes in that internal configuration and what is actually going on up there in the sky as the planes move around?"

The answer is that there seems to be no reason why that link has to be of any one particular form. The fact that up until now the link adopted has been a fairly direct (though nonetheless still pretty ethereal) one is simply because the entire computing field is still very much in its infancy. By and large the present practice is to use individual locations (computer "words") to represent objects in the world (airplanes, say) and to impose electrical connections between these locations in a manner that corresponds to things going on in the world, using mathematical logic (in some form or other) as the main tool.

For example, if airplane A's existence on a particular flight path means that airplane B should remain on the ground, the computer program will provide an appropriate link between those memory locations that store all the information about A and those that relate to B.

This approach amounts to regarding particular areas (i.e., collections of locations) within the computer memory as corresponding to particular objects (or parts of the real world), with circuit links between these memory areas corresponding to the various activities that go on between those objects.

It is an approach that has a number of attractions. For a start it is a very obvious one. It allows the computer programmer to visualize what is going on inside the machine in an extremely helpful manner. It makes for relatively easy modifications to the program. (If another aircraft enters the area, just allocate an unused area of memory to represent it in the same way as all the others and link it in the appropriate fashion.)

The only drawback to this approach is that for most of the really interesting (nonmathematical) problems for which people want to use computers today, it simply does not work. It falls victim to what is known in the trade as "combinatorial explosion." Though in theory the computer can solve your problem, what happens in practice is that the computation requires so many steps that, for all their speed, even the most powerful of today's machines would require millions of years to come up with an answer.

The solution? Quite simply, to start all over again. Not necessarily (though maybe) with the way we build computers, but with the way we use them to store and manipulate information about the world.

The key lies not so much in what goes on inside the machine as in the manner in which that internal activity relates to the world it is supposed to represent. Grasping that key will most likely require the development of entire new areas of mathematics and engineering. *Micromaths* intends to start off the new year with a round-up of some of the new work currently being done towards this goal.

- *Time*'s Planet of the Year:
  Endangered Earth
- USSR withdraws from Afghanistan
- Benazir Bhutto: First woman to lead
  Muslim state
- Fires char 995,000 acres of
  Yellowstone
- Air show disaster in West
  Germany kills 70
- Aloha Airlines 737 lost huge
  chunk of roof but no
  passengers
- Michael Dukakis for
  President
- Jesse Jackson
  for President
- Iran uses poison gas on Kurds
- Olympic Games: Seoul and Calgary
- Jimmy Swaggart confesses sins
- Battling Tysons
- USS Vincennes downs Iranian
  passenger jet
- Robert Morris's virus kills 6,000
  computer systems
- Steffi Graf wins Grand Slam
- Bush elected President

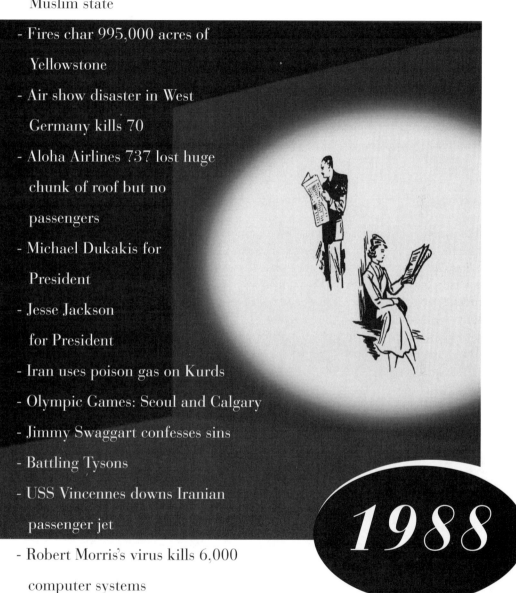

1988

101

Game, set, and match program
January 7, 1988

When Wimbledon tennis champion Pat Cash returns a particularly vicious first service, his movements are so fast that it is often difficult for spectators to follow his actions with any accuracy. How does he do it? And, more relevant to readers of *Computer Guardian,* is there any way of building a machine that could do as well?

Leaving aside the nuts and bolts engineering problems involved (and they are considerable), what would be the best way to proceed as far as designing the all important software control system—the "brain" that directs the actions of the device?

The traditional artificial intelligence approach would go something like this. Presumably the robotic tennis player has some form of visual apparatus. By dividing up the field of vision with a geometric grid, the position of the ball at any moment can be stipulated in terms of numerical coordinates. By recording these coordinates over a number of fixed intervals of time as the ball begins its flight, some simple mathematical computations will give both the speed and the direction of flight.

Some further mathematics and a bit of high school physics can then be applied in order to produce a mathematical equation that predicts the exact path the ball will follow until its flight is over. Solving this equation will then produce the coordinates that the ball will have the moment it comes within range of the robot's racket. By writing a computer program to automatically perform each of these computations, the output from this program can be fed directly to the appropriate mechanism controlling the robot's arm.

Then a further set of equations must be solved in order to maneuver the racket into exactly the right position so that it hits the ball. Or at least, so the ball hits the racket. There is another problem that has not even been touched upon in the discussion so far. How does the robot decide where to try to direct the return, and how is the return then achieved?

Even assuming the simplest case of returning the ball directly to the server, a further set of equations (involving more physics, knowledge of gravity, the elasticity of the ball, laws of momentum, and so forth) have to be formulated and solved in order that the racket strikes the ball with the correct force and at the appropriate angle.

So how does Pat Cash manage it? In his case the problems are even greater than for our hypothetical robot. He not only keeps his eye on the ball and positions his racket accordingly, he moves his entire body, chooses the best place to target the return, and even anticipates where the server might direct his second shot.

Leaving aside Mr. Cash's mathematical abilities (about which I know nothing), it seems obvious that he does not win Wimbledon titles by solving differential equations rapidly in his head. In which case there must be some other way to solve the problem. What way? And more to the point, why don't the AI people adopt this other approach?

The second question is far easier to answer than is the first. Current practice in AI follows more or less traditional mathematical approaches to problem solving because that is the only way anyone knows how to proceed. What is more, given the right task, such a procedure works. It did, after all, get a man to the moon and back.

Where the method falls down most spectacularly is when it is applied to just those tasks that people perform with the most ease: tasks such as playing tennis (though not always as skilfully as Pat Cash), recognizing pictures and scenes, and understanding everyday language.

Overcoming the difficulties presented by problems of this sort would appear to require a quite different approach to computing itself. An approach that will involve both new mathematics and new engineering techniques. Namely, the design and construction of computers that operate in a manner much more closely akin to the way the human brain functions.

A significant pointer in this direction comes out of some work centered at the University of California at San Diego, and described in the two-volume book *Parallel Distributed Processing,* by David Rumelhart and James McClelland (Addison-Wesley, 1987). Though the work of the San Diego group was motivated primarily by the desire to model (using computer simulations) human learning and thought processes (for psychological purposes), the implications for the future of computing are clear.

Briefly, what the group did was this. Using a traditional computer as a simulation tool, they designed and built a quite new kind of computation device, known as a "neural net." This consists of a large number of very simple computational devices, all connected together to form a complex network. Each simple computational unit is able to decide which of the other units it will "talk to" at any moment, and to what extent it will exchange information with those other units. But that is all. No stored database. No carefully worked out program.

The only principle that guides the machine is that its design incorporates notions of "good" and "bad," and it is constructed to try to be good. When the machine is first switched on, there follows a period of initial learning, highly reminiscent of the way parents instruct their children. Each input to the machine produces an output. If it produces an unsuitable output, the machine is informed of that also.

By adjusting its internal communication network, the machine strives to increase the number of "good" results. At first by blind trial and error. Later on, as the learning process continues, by a mixture of trial and error and experience. Eventually, the machine behaves exactly as if it "knew" whatever it was the instructor was trying to tell it.

By this stage in the process, the instructor (he cannot really be called a programmer any more) does not know exactly what is going on inside the machine—it is far too complex for that. But the behavior of the device measured in terms of what output results from what input becomes more and more like intelligent (or skillful, depending on the task) activity.

And that might well be the way computing goes in the near future. The computer programmer will be replaced by the engineer and the instructor. We will no longer understand exactly how our machines work, but they will work far better than the ones around today, and we shall have to be content with that.

The next *Micromaths* will look in more detail at one of the "neural net" computing devices developed by the San Diego group. [See Chapter 102.]

102

Doing it the brain's way

January 21, 1988

Hailed as the all-powerful information processing device of the future, the electronic computer has proved stubbornly difficult to tame when it comes to many tasks that people perform with ease. Give it a complicated piece of arithmetic, or a mathematical equation to solve, and the humble home micro will perform brilliantly. But tasks such as recognizing a simple scene or dealing with everyday language cause the most powerful mainframe to struggle—and, more often than not, to fail.

Why is this? For people, and indeed for most living creatures, the exact opposite is the case. Though a game of chess presents most of us with a considerable mental challenge, we have no difficulty in finding our way across the room without falling over or bumping into things. We do it without thinking.

And there in a single sentence may be found the answer to our question. Or so, at least, a growing number of workers in Artificial Intelligence (AI) are beginning to say. According to them, the reason it has been so difficult to program computers to perform simple tasks of scene recognition and robot control is that to consider these processes as a series of steps that can form the bases of a computer program is just not the right way to do the job. This is certainly not the way the human brain deals with such tasks. Given the much slower operating speed of the individual neurons that make up the brain's compu-

tational power—silicon is some 10,000 times faster—the brain has to solve any "real time" task in at most 100 steps or so.

Instead of raw computing speed, the brain relies upon its massively parallel architecture to get the job done on time. That much is known. Just how it uses this parallelism is less certain and the subject of some controversy. Also controversial (though of late receiving an increasing share of the computing funding cake for all that) is the question of whether there is any real chance of building machines that operate in a brain-like fashion. But for a real, good old-fashioned bare-knuckles controversy, you have to look at the way the brain-computer analogy is used in the other direction, to provide evidence for the way we human beings think.

In 1986, David Rumelhart (then at the University of California at San Diego, now at Stanford University) and James McClelland (at Carnegie-Mellon University in Pennsylvania) published their findings concerning a computer model of the way in which the past tense of English verbs may be learned. They thereby started off a fierce debate that has so far spilled over from the halls of academe and into the pages of the *New York Times* and magazines such as *Science*.

The Rumelhart–McClelland model simulates on a normal computer a so-called "neural network," a computational system consisting of a large number of simple computational units connected together to form a complex network. Such systems are by no means new. Research groups both at San Diego and Boston University, among others, have carried out extensive investigations into their operation. (See for example the two-volume work *Parallel Distributed Processing* by McClelland and Rumelhart, published by the MIT Press in 1987.)

What caused the controversy in the case of the past tense work was the way the computer seemed to acquire its knowledge. The way it learned was highly reminiscent of the manner in which young children learn to use the past tense. With a computer behaving like our own children, making the same mistakes and then correcting them in the same way, it was inevitable that a huge debate would spring up as

to just how significant this development was in psychological terms.

In a recent lengthy paper, Steven Pinker of MIT and Alan Prince of Brandeis University subject the Rumelhart–McClelland model to an unusually detailed scrutiny, pointing to a number of reasons why the model does not, in their view, substantiate the claims made by its developers. An attack that Rumelhart, in a talk given at Stanford late last year, said was considerably off target, concentrating on technical details of the model and thereby missing its import as a first pointer to things to come. The debate has clearly only just begun.

What makes the whole affair so fascinating is that the issue could well affect the entire future of computers and the way we build and operate them, as well as our understanding of the way human beings think and learn.

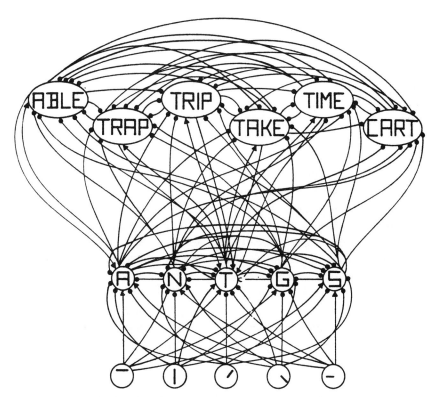

FIGURE 32.   Neural computing. A simple neural network.

The idea behind a neural network computer is a simple one. A large number of extremely simple computational "units" are all connected together. Each unit consists of a switch that at any time can be one of a fixed number of positions. The position of each switch is determined by the inputs it receives from all the other units, and in turn determines the output from that switch.

The signal sent from any one unit to any other has its strength modified along the way by the "connection strength" of that particular connection. The connection strength may itself be modified by the positions of the switches at each end. And apart from input and output, that is all there is to it. No separate memory or processing units. No database. No list of instructions. The entire system just runs itself as a single entity.

In such a system it makes no sense to point to a particular location (unit) and say that this represents some particular data item, or to indicate some subcollection of units and claim that this is where a certain logical operation takes place.

For such a machine, programming is replaced by a form of "teaching." The machine starts off "knowing" nothing. But by repeatedly giving the machine selected inputs (particular scenes or geometric shapes if the task is scene recognition, words or sentences if the aim is to handle everyday language), and informing the machine when it produces the correct output response, the machine can be "trained" so that it progresses from initially producing totally inept responses to eventually performing with considerable (and oft unnerving) accuracy.

For their past tense model, Rumelhart and McClelland took a particularly simple network consisting of two banks of 460 units, each unit in one bank connected to every unit in the other. One bank accepted input, the other produced output. Input took the form of English verbs presented in present tense. Output was intended to be the corresponding past tense of that verb. No hidden database. No look-up table of past tenses. No built-in rules for forming past tenses. The intention was that the machine should produce the correct past tense of a given verb in a single machine cycle (i.e., input gives rise at once to the corresponding output).

Since Rumelhart and McClelland were interested in the phonetics of language as well as the mere spellings of words, verbs were presented to the input units in the form of a whole range of parameters known as "Wickelfeatures" that combined to give all the information about the various words that were of interest. Output consisted of a set of Wickelfeatures too.

Using Wickelfeatures it is, for example, possible to distinguish between the three ways of pronouncing the 'ed' ending on past tenses exemplified by the past tenses 'looked' (a T sound), 'lived' (a D sound), and 'melted' (an ED sound).

With their model thus set up, Rumelhart and McClelland went on to train it to produce past tenses of verbs. They took a collection of some 420 verbs, chosen to include many troublesome (for humans) irregular verbs as well as more standard regular verbs, and trained the machine with 200 training cycles, whereby each verb was fed in in its present tense and the output compared with the correct past tense. (The result of the comparison then being fed back into the computer so that it could modify its future behavior).

And this is where, to their initial surprise, Rumelhart and McClelland observed that their "baby" was behaving just like children do when they learn past tense in English. At first the machine's responses were totally erratic. It clearly "knew" nothing about past tense formation. Then after a while it began to be successful with some verbs, including such difficult irregular verbs as *hit, make, sing,* and *go.*

But then it acted as if it had noticed that the majority of verbs behave in a very regular manner, forming their past tenses by the addition of 'd' or 'ed,' and the result was that the machine started to over-regularize, producing such past tenses as *hited, maked, singed,* and *goed.* Only subsequently, after several more training cycles, did the machine learn to distinguish the regular verbs from the irregular ones, and to distinguish the different forms of pronunciation that the regular past tense can involve.

This is precisely the same learning process that has been observed in small children learning English. And thereby comes the controversy. For conventional theory has it that children learn language by gradually picking up "rules" which they then apply at first indiscriminately (over-regularization), and then in a more sophisticated fashion. Quite in line with the accepted wisdom in AI, in fact (socalled rule-based systems).

But in Rumelhart and McClelland's machine there is no way in which a "rule" may be learnt and then applied. Rather it simply adapts its crude input-to output-reaction in a way that at most may be said to be *in accordance with* (as opposed to obeying) such a rule. But it does appear to work.

Is this then a pointer to the fact that much human activity is nothing more than simple action-reaction? Possibly.

Does it mean that current attempts to solve recognition and language problems using traditional AI methods are doomed to fail? Again, possibly.

Are we witnessing the first step towards a quite new way of building computers? Well, regardless of the outcome of the present debate, my guess is that if cruise missiles are still around in ten years time, they will be guided with the aid of a neural-net computer.

103

Mud on whose face?

February 4, 1988

Recently, it began to look as though the entire North American computing fraternity was about to grind to a halt as people from San Diego to Boston wrestled with a simple probability question.

It all started innocuously enough, when someone submitted to one of the many public electronic bulletin board systems (BBS) this little teaser. Two coins are placed in a drawer. Each coin is either gold or silver, but they are not both gold. Apart from that you know nothing about the two coins. Without looking, you now reach into the drawer and take out a coin. It turns out to be silver. The question now is this. All things being equal, what is the probability that the remaining coin is silver?

Sound easy enough? Most people think so. The standard response runs like this. There is one coin remaining in the drawer. Since the question says that you should assume "all things are equal," there is a 50-50 chance that this second coin is silver as opposed to gold. So the answer is 1/2.

Unfortunately, this is not the right answer. Your task as a *Micromaths* reader is to figure out why not, and to come up with the correct solution to the problem.

At this point I should warn you that this problem has a well-established track record of causing anguish. That computer bulletin board I mentioned groaned for weeks under the weight of endless discussions of just this one topic, and at times it seemed that only the often vast distances separating the participants in this modern day electronic debate kept them from coming to blows. [See Chapters 104 and 106.]

Incidentally, there is no catch to the question. It is a straightforward matter of figuring the correct odds. Only it turns out to be not quite so straightforward after all.

And if that is not enough, see what you make of this one, that formed the basis of some discussions at Stanford University last month.

Claire and Max are playing in the garden and both get mud on their forehead. Neither knows that they have a dirty face, but each can see the mud on the other. Since their father has warned them against getting dirty, each is secretly pleased that the other has a muddy forehead while they themselves are clean (they think).

Father comes out and says sternly "Oh no, at least one of you has mud on your forehead."

At first nothing happens, as the two children continue to smile smugly at each other. Then suddenly, each child realizes that they have a muddy face and they both start to cry.

The first question to ask is, how did the children manage to figure out that they had mud on their own face? (You have to assume that the kids are pretty smart, but then this is a Stanford story). The crucial, and only, clue the children had, was that their father told them at least one of them had a muddy face.

Now comes the tricky part. Without a doubt, it was what the father said that enabled the two kids to figure out that both their faces were dirty. Before then, each could see that the other child had a muddy forehead, but was convinced that they themselves were clean. But wait a minute. In telling them that he could see at least one dirty face, the father was surely only informing them of something they knew already. In which case, how come this made any difference at all to the situation?

The only clue I shall give is that the answer is connected with problems in economics, politics, psychology, nuclear deterrence, gambling, and computer science.

But does this clue help the *Micromaths* reader?

104

The silver coin tease

February 18, 1988

As I expected, the coins in the drawer teaser [Chapter 103] caused no end of debate—just as it had when it appeared on the North American electronic bulletin board from which I took it. Here briefly is the problem.

A drawer contains two coins, each one either gold or silver. You do not see the coins, but an observer tells you they are not both gold. Beyond this you know nothing (and must therefore assume all possibilities equally likely). You take one coin out of the drawer at random and it turns out to be silver. What then do you judge to be the probability that the other is also silver? The most popular solution was 1/3, with 40% of respondents plumping for this figure, closely followed by a 35% vote in favor of 1/2.

After these two popular solutions, the replies split as follows: 15% thought 2/5 was the answer, 5% went for 1/4. Everyone who wrote presented a clear (and in most cases convincing) argument to support their result. This demonstrated quite clearly that questions about probabilities, when posed in loose, everyday language as in this case, have an inherent ambiguity. After a quick summary of the two most common solutions, I'll tell you which one I think is the right answer.

Argument 1 runs like this. As the possibility of two golds is ruled out, there are three equally likely ways of picking out the two coins: GS, SG, SS. Only the last of these gives silver followed by silver, so the probability of this happening is 1/3.

Argument 2 is a mathematically sophisticated one that goes like this. The probability of the drawer containing two silver coins (possibility A, say) is 1/3, the probability of a gold-silver combination (possibility B) therefore being 2/3. Then, using standard notation for conditional probabilities, $p(S \mid A)$, the probability of picking a silver coin (event S) given A, is equal to 1, and $p(S \mid B)$, the probability of picking a silver coin given B, is 1/2. What the question asks for is the conditional probability $p(A \mid S)$. This may be obtained from the figures just stated using a standard result known as Bayes' Theorem:

$$p(A \mid S) = \frac{p(A) \cdot p(S \mid A)}{p(A) \cdot p(S \mid A) + p(B) \cdot p(S \mid B)}$$

Substituting in the various values on the right-hand side of this equation gives the answer $p(A \mid S) = 1/2$.

So are either of these correct? Well, it all depends on what you think the question is asking for. Of these two solutions, my vote definitely goes for 1/2.

The problem with the answer of 1/3, I think, is that it does not deal with the situation you are faced with after the first coin has been chosen—a rather different situation altogether.

But in point of fact, I don't think that either of these figures answers the question being asked. According to my reading of the situation (which, after all, I wrote, and worded rather carefully I might add), having ruled out a gold-gold combination, there are two equally likely possibilities: a gold-silver combination or a silver-silver pair. (In each case you can look at the pair either way round, of course, though in the case of the silver-silver combination you will be unable to tell the two orderings apart.) Picking out a silver coin in the silver-silver case is clearly twice as likely as in the silver-gold case (a probability of 1 against a probability of $1/2$). So I would figure a $2/3$ chance that the second coin was also silver.

But it does all depend upon how you interpret the initial information, and what it is you think you are calculating.

Incidentally, in the electronic bulletin board debate that sparked off this whole issue the participants seemed to split fairly evenly between the figures $1/2$ and $2/3$. The *Micromaths* readers' favorite of $1/3$ was hardly mentioned, though I don't know what this means.

Perhaps the safest answer was provided by those readers who simply declared the problem to be insufficiently precise. [See also Chapter 106.]

105

Prime the record books
March 10, 1988

The recent discovery of a new Mersenne prime number has not only focused attention once again upon the immense power of modern computers, it has also raised questions as to whether any other such numbers have been missed in the hunt to get into the record books. For this one, the 31st ever to be discovered, is smaller than the two largest known Mersenne primes, and was found in a region long thought to be devoid of such numbers.

The development provides one of those few occasions where anyone can catch a glimpse of the fascinating, but largely impenetrable, world of pure mathematics. For one of the beauties of the prime numbers lies in their extremely simple definition, a prime number being a positive whole number that can only be exactly divided by itself and by 1 (examples are 2, 3, 5, 7, 11, 17, 19, the only primes less than 20).

And yet, for all their simplicity, the prime numbers are of immense importance in mathematics, being to the mathematician what the elements are to the chemist or the fundamental atomic particles to the physicist. Prime numbers are the stuff out of which all other numbers are made, and as such their study is both intricate and deep, having enormous implications both for mathematics itself and for applications of mathematics.

But the deep implications do not rival the straightforward news-value of an announcement that a new world record prime has been discovered. The first couple to hit the headlines in this way were two 18-year-old American high school students, Laura Nickel and Curt Noll, who in 1978 used a Cyber 174 computer at the California State University at Hayward to discover a record prime number having 6,533 digits.

The current record holder is a 65,050-digit prime number discovered using a Cray XMP supercomputer in September 1985. [Chapter 63.] But in their hunt for world records, there has been a tendency to take some short cuts that have left some unanswered questions, and left the way open for the latest discovery.

Identifying a large prime number is essentially a computing problem. Just how do you go about establishing that a number having thousands of digits has no exact divisors other than itself or 1?

Checking all smaller numbers (or even all primes less that the square root, which is a more sophisticated approach) simply will not work. For a number having a mere 50 digits it would take even the most powerful of today's supercomputers a staggering 10 billion years to check for primality in this way. So more ingenious methods have to be employed, often using quite advanced mathematical ideas.

The most successful approach, as far as finding record primes is concerned, has been to look for *Mersenne primes*. A *Mersenne number* (named after a seventeenth-century French monk of that name who first investigated primes of this form) is one of the form $2^n - 1$ for some number n. For such numbers, there is a special test for primality (known as the Lucas–Lehmer Test) that works very fast. And because of the speed at which Mersenne numbers grow to be extremely large as you increase the value of the exponent n (a stack of 2^{65} ten-pence pieces would reach Proxima Centauri, the nearest star to the solar system), these numbers therefore provide an obvious hunting ground for record primes.

The question of mathematical interest is just how many Mersenne numbers are there that are prime? Here Euclid's proof does not help. No one knows as yet whether there are infinitely many Mersenne primes or not. In fact, until the latest discovery, only 30 such numbers had been discovered, the largest two being the 39,751-digit Mersenne prime $2^{132049} - 1$ (discovered in September 1983) and the current record holder $2^{216091} - 1$ mentioned earlier. [See Chapters 1, 2, 4, 8, 63, and 130.]

But in the rush to get into the record books, the prime number hunters had started to take chances, leaping up to ever higher exponents rather than checking each one in turn. There was always the chance that a Mersenne prime would be overlooked in the process, though statistical evidence was used to try to avoid this possibility, by trying to predict the approximate size of those exponents n for which the Mersenne number $2^n - 1$ is prime.

Now, it seems, this has in fact occurred. Walter Colquitt and Luther Welsh, two workers based in Houston, Texas, have recently discovered a 31st Mersenne prime. Their number has the exponent 110,503, which makes it smaller than the two record holders mentioned above. Besides throwing into doubt some of the statistical predictions used in earlier work [see Chapter 4], this new prime opens up the possibility of further discoveries of "lost Mersenne primes."

Postscript.   See also Chapters 1, 2, 4, 8, 63, and 130.

106

Silver coins and gold in the box
March 24, 1988

The coins-in-the-drawer problem from February 4 and February 18 [Chapters 103 and 104] continues to bring in a steady stream of mail, a lot of it objecting to my proposed solution given on February 18. Here briefly (and definitely for the last time) is what it said.

Consider two coins hidden in a drawer, each gold or silver, not both gold. A coin is taken out at random. It is silver. What is the probability that the other is silver?

The most common answers people come up with are 1/3, 1/2, and 2/3. Of these, my choice is 2/3, closely followed by 1/2, with 1/3 totally rejected. However, the answer of 1/2 is definitely not justifiable on the grounds that it is like tossing a coin twice. With the coins-in-drawer set-up, the first choice of a coin very definitely does affect the final outcome (unlike tossing a coin twice, where the two results are quite independent).

Beyond that, however, it is hard to say just what is "right" and what is "wrong."

The point is that *as stated,* the problem is not purely a mathematical one. It is more a question of what is known as "subjective probability": faced with a situation exactly as described, how would a rational person (in this case you) act?

To obtain a *definitive answer,* you would need to know exactly what procedure was followed in getting the coins into the drawer in the first place. In the absence of that information, you simply have to make a subjective judgment, which results in different people coming up with different answers.

And here you have an inkling of why this particular problem is in the computer section. One of the problems facing people who try to build intelligent robots, or other forms of "artificial intelligence" systems, is how do you get a machine to act in a "commonsense" manner? Give a computer all the information it requires, and it will come up with the right answer. But in most real-life situations, complete information simply is not available. Decisions have to be made using just what does happen to be known.

Trying to figure out how to get machines to make such judgments is proving to be no easy task. Probability theory was at one time thought to be the key, but as problems like the coins in the drawer indicate, the apparent certainty given by results such as Bayes' Theorem (a favorite method of solution for many *Micromaths* readers) can be misleading, since its mathematical sophistication tends to mask the fact that its use may depend on some assumptions for which there is no justification at all. In this case, assumptions about how the coins were first put into the drawer. All that can be said is that on the basis of what was stated in the problem, it was likely that the initial set-up was such-and-such.

But as the results of this *Micromaths* episode demonstrate quite clearly, one man's common sense is another's "You are clearly wrong."

Though folk will doubtless continue to investigate "common sense reasoning" and "reasoning under uncertainty," the man in the street (and even more the man in the computerized doctor's consulting room) should be very wary about what comes out of such systems. The silicon-chippery and the formidable-looking mathematics might seem very impressive, but could well be masking what are some very dubious looking assumptions indeed.

Meanwhile, I can't resist throwing in another teaser of a similar, though far less controversial nature.

You are faced with three boxes, in one of which is a pot of gold. (The other two boxes are empty.) What you will be asked to do is the following. First pick a box. After you have done that, one of the two remaining boxes will be opened (by the person who set up the game, and who therefore knows the contents of each box) to reveal that it is empty. You will then be allowed either to stick with your original choice or to change it.

So what strategy would you adopt: sticking or switching?

Most people initially assume that it cannot possibly make any difference. They might just as well stick with their original choice. But this is quite definitely wrong. If the switching tactic is adopted, you double your chances of picking the pot of gold.

Just why this is, I shall leave you to figure out for yourself.

107

The security in big numbers

April 21, 1988

At the end of March, Robert Silverman, an employee of the Mitre Corporation in Bedford, Massachusetts, announced the first ever factorization of a 90-digit number using a general purpose factorization technique.

The number he factorized is arrived at by first raising 5 to the power 160 and adding 1, and then dividing by the result of raising 5 to the power 32 and adding 1, that is:

$$(5^{160} + 1)/(5^{32} + 1)$$

The principle interest in this particular number is that it has for some time been recognized as being particularly resistant to a whole range of modern factorization algorithms.

A great deal of the present-day concern with methods for factorizing numbers stems from the discovery, just over a decade ago, of a new form of data encryption system that depends for its security upon the practical infeasibility of factorizing numbers with 100 digits or more. [See Chapters 2 and 13.] This development has, on occasion, led to attempts to hide research on factorization beneath the cloak of official secrecy. The discovery of an efficient method for factorizing would at once render this security system useless.

But much of the fascination lies in the simple and basic nature of the subject. The positive whole numbers are the most fundamental of all mathematical objects. Positive whole numbers fall into two important categories:

the *primes,* those numbers that can only be evenly divided by themselves and 1, and the non-primes, or *composite* numbers. For example, each of the numbers we 2, 3, 11, 19 is prime, while the numbers 4, 6, 12, 30 are composite.

Mathematicians have devised very efficient methods for testing (using a computer) whether numbers with up to a hundred or more digits are prime or composite, but these methods do not work by looking to see if the given number has a proper divisor or not. (That approach could take billions of years just to check that a single 50-digit number was prime.)

What has proved altogether more difficult has been to devise an efficient method that would find a proper divisor, or factor, of a number that had been shown to be composite. Indeed, it is precisely this distinction between the relative ease of testing for primality and the seeming impossibility of factorizing that lies at the heart of those cryptographic applications mentioned a moment ago.

Some of the more recent factorization work has concentrated on the development of so-called "Monte Carlo methods," that depend for their success on probabilistic techniques. As the name aptly suggests, if luck is on your side, these methods can be extremely rewarding, and numbers with a hundred digits or more have been factorized. But not all numbers can be factorized in this way. In particular, the kinds of numbers used in cryptography are not in general amenable to such an attack.

What general purpose, non-probabilistic methods there are tend to fall down on numbers having more than about 85 digits. This is true even if the computation is carried out using one of the most powerful supercomputers available, such as the machines produced by Cray Research.

So how did Silverman manage to factor his 90-digit monster? There were two significant features that made it possible. First, he used one of the best general purpose factorization techniques around, the so-called "Multiple Polynomial Quadratic Sieve." When implemented on a supercomputer, this method can routinely factor numbers with 85 or so digits in a couple of hours.

Second, he used a massive amount of computer time: not a single supercomputer, but something that proved to be just as good, and in some ways better. For his attack, Silverman assembled a parallel network of two dozen Sun-3 workstations. The entire network worked on the problem for some 625 hours, giving a total computation time of something like 15,000 hours, or one year and eight months.

This sounds a lot, but a previous attempt to factorize this particular number had lasted five years without success!

The two factors found by Silverman's program have 41 and 49 digits respectively, which provides a second world record. The smaller factor is the largest penultimate factor ever found of any number.

So what is the significance of the new factorization as far as the security business is concerned? The answer is, actually, very little. The length of time required to complete the job is far too long to present a threat to today's sophisticated cryptographer, and there was no new mathematics involved.

But it does serve to emphasize once again that progress is being made. For without the modern algorithm and sophisticated programming techniques Silverman used, no amount of computing power would have resulted in success.

108

The deadly traps in simple problems

May 5, 1988

The recent rise, and rapid fall, of yet another attempt to prove Fermat's Last Theorem (*Micromaths,* April 7 [not included in this compilation]) emphasized once again just how easy it can be to pose simple looking questions whose solutions turn out to be very far from easy.

When he wrote his now famous note in the margin of a textbook to the effect that the equation:

$$x^n + y^n = z^n$$

has no genuine whole number solutions for the unknowns x, y, z whenever the exponent n is greater than 2, the 17th century mathematician Pierre de Fermat thought he was able to prove this result. What evidence there is suggests he may have had a proof for the case $n = 4$, and just conceivably for the case $n = 3$ as well. But it is highly unlikely that he had proved the full result he claimed.

More likely he fell into one of several subtle, but ultimately deadly, traps that await the would-be Last Theorem prover. Just as did Leonhard Euler, the great Swiss mathematician, who published a "proof" for the $n = 3$ case in 1770. Though fiendishly clever, and ultimately rescuable (though only by virtue of an accidental aspect of his argument), Euler's reasoning depended on what was probably the same invalid, key step that led Fermat and numerous other highly gifted mathematicians

astray. Ernst Eduard Kummer developed a means of getting round the difficulty, but only by developing an entire new branch of mathematics, nowadays called Ideal Theory.

Building upon Kummer's work, modern computer technology has been brought to bear on the Last Theorem, so that at present Fermat's claim is known to be true for all exponents n up to 150,000. But that falls far short of the exact proof required by mathematics.

In fact, until recently there was no convincing evidence to suggest that the Last Theorem might be true, other than the fact that it was known to hold for those first 150,000 values of the exponent. Then, last year, it was shown that the Last Theorem follows from the so-called Taniyama–Weil conjecture which, though itself not proved, at least has some impressive evidence in its favor.

Until that result, there was really no more to recommend the Last Theorem than there was to support the Euler Conjecture that the four variable equation

$$w^4 + x^4 + y^4 = z^4$$

has no genuine whole number solutions.

This conjecture remained unresolved for two centuries until, last summer, the Harvard mathematician Noam Elkies proved that it was false.

Elkies' solution involved some enormously large numbers, discovered by a combination of mathematical theory and computation. After a long computer search, Roger Frye of the Thinking Machines Corporation, discovered the more manageable solution

$$w = 95,800, \qquad x = 217,519,$$
$$y = 414,560, \qquad z = 422,481.$$

For all we know, the evidence from last year notwithstanding, the same fate may await Fermat's Last Theorem.

Then there is the long standing Twin Prime Conjecture. A pair of prime numbers is said to be a *twin pair* if they are just 2 apart: for instance, 3 and 5; 17 and 19; 227 and 229. It is conjectured, but not proved, that there are infinitely many such twin pairs, just as there are infinitely many prime numbers altogether. The largest such pair known consists of two prime numbers with 2,259 digits: the two numbers you get by respectively adding and subtracting 1 from the even number consisting of 107,570,463 followed by 2,250 zeros. This particular pair was discovered by Harvey Dubner in 1986, using a homemade computer specially designed for handling large whole numbers.

Still another innocuous-looking teaser is the Goldbach Conjecture. In a letter written to Euler in 1742, Christian Goldbach suggested that every even number greater than two can be written as a sum of two primes. For instance, $4 = 2 + 2$, $8 = 3 + 5$, $20 = 3 + 17$. Try a few more values, and you will see that Goldbach's conjecture certainly seems to be true. And if you write a computer program to carry the search still further, you will still obtain nothing but collaboration for the conjecture. In fact, computer searchers have been taken all the way up to a billion and beyond, still without coming across a single even number for which it fails. But for all that, there is still no reason to believe that the result is true.

Certainly, it is unwise to base a firm conclusion on numerical evidence, no matter how "overwhelming" it might seem. Take the case of the Mertens Conjecture [Chapter 19]. This long-standing conjecture about numbers was demonstrated to be true for all whole numbers up to 10 billion before it was shown to be false in 1983.

Simplicity, then, is no guide to ease of solution when it comes to questions about whole numbers.

109

Hunt goes on for maximum factors

May 19, 1988

Within weeks of a new record being established in the computer factorization of large whole numbers, that record has itself been surpassed.

As reported in *Micromaths* on April 21 [Chapter 107], the end of March saw the first ever factorization of a 90-digit number using a general purpose factorization technique. Robert Silverman of the Mitre Corporation in Bedford, Massachusetts, ran a network of two dozen Sun-3 workstations for a total of 625 hours in order to find the 41- and 49-digit prime numbers that when multiplied together yield the 90-digit number:

$$(5^{160} + 1)/(5^{32} + 1)$$

a number whose prime factors had long resisted detection.

The new record involves the factorization of a 92-digit number. Given that only a few years ago the factorization of even 50-digit numbers was regarded as a significant achievement, this latest development indicates the tremendous advances that have been made, and doubtless will continue to be made, in this peculiar hybrid field, that lies somewhere between classical mathematics, computer science, and engineering.

In fact it is the cooperative interplay between the theory and the engineering that is one of the most fascinating aspects of this kind of work. (The other motivation is provided by the relevance of factorization to modern cryptographic systems, described on several occasions in this column.)

For the average pure mathematician, it is usually enough to know, as did Euclid in 350 B.C., the raw fact that every positive whole number is either a prime number (that is, has only itself and 1 as exact divisors) or else factors into a unique product of prime numbers. Knowledge of the actual prime numbers involved for any particular number is rarely of any importance.

Indeed, given the infinitude of the whole numbers, even those dedicated mathematicians who pursue factorization as their principal research project are not interested in what the answer turns out to be—the sheer size of the numbers involved precludes that. Rather the hunt is carried out much more in the spirit of mountain climbing. The thrill and the challenge are provided by the search itself.

To provide the would-be factorer with a suitable goal, there is even a published list of numbers to try to factorize: Wagstaff's List. It consists of nothing more than page after page of numbers of the form $b^n + 1$ and $b^n - 1$, together with whatever is known of their factorizations. But notwithstanding the lack of a good plot, Wagstaff's book is already into its second edition, and the latest update is due out some time this summer.

Like the police lists of wanted criminals which it resembles, Wagstaff's book classifies the numbers it contains under headings such as "more wanted" and "most wanted." The classifications are assigned largely as a result of the difficulty the number seems to present with regards to factorization.

The number $6^{131} - 1$ is one such "most wanted" number. Part of its prime factorization was already known. This number is divisible by each of the primes 5263, 3931, and 6551. That much is easy to discover. Anyone with a home micro can write a program to work out the number $6^{131} - 1$ and search for all prime factors with only four or five digits. It is the number left when the original number is divided out by these four small factors that proves to be the real hard nut to crack. This is the 92-digit number that has just been factored. The record-breaking factorization took factoring experts Herman te Riele, Walter Lioen, and Dik Winter of the Centre of Mathematics and Computer Science in Amsterdam some 95 hours of computation on the NEC SX-2 supercomputer of the Dutch National Aerospace Laboratory in the Netherlands.

Besides highlighting the power of the machine used, which is generally regarded as the world's fastest single-processor vector computer, the new result again emphasizes the potential power of the factorization technique used. Known as a Quadratic Sieve Method, this technique is a so-called general purpose algorithm that will work on any number, not just numbers of certain special forms, as is the case with some other methods that have been used to factor numbers with over 100 digits.

Moreover it is a technique that, in its essentials, goes back to the number theory of the 17th Century amateur mathematician Pierre de Fermat. This provides the great Frenchman with a second reason for laughing from his grave within a matter of months. No sooner has his famous Last Theorem defeated yet another attempt at solution [Chapter 108], than his methods have been used to break the world factorization record.

The factorization also establishes a second new record: the 92-digit number is the largest to be factored using a single processor computer. The previous record here was an 82-digit number, factored by the same team in December 1986 using a CDC Cyber 205 supercomputer. The new factoring program on the NEC SX-2 runs about 10 times as fast as did its predecessor on the Cyber.

There are some things you simply cannot do on a PC.

110

Beauty figures

June 2, 1988

British mathematician G. H. Hardy said it all in his 1940 book *A Mathematician's Apology*, "The mathematician's patterns, like the painter's or the poet's, must be beautiful, the ideas, like the colours or the words, must fit together in a harmonious way."

Two items this week testify to this claim. They are items that Hardy would have had difficulty investigating, but which today's home micro bring within the reach of anyone able to write simple programs.

The first concerns numerical beauty: the pleasing symmetry exhibited by palindromic numbers (numbers that read the same backward as they do forward) such as 101 or 12321. (Constant digit examples such as 999 are not generally regarded as palindromic.)

The task facing the home micro owner is to find palindromic numbers whose square is palindromic. For example, 121 is a palindromic number that squares to the palindromic number 14641; the palindrome 202 squares to 40804; and 20102 squares to the palindrome 404090404.

If you carry your search far enough, you will discover some extremely attractive results. For instance, when you square the nine-digit palindrome 101010101 you get the number 10203040504030201, which is just the digits from 1 up to 5 then back down to 1 again, each separated by a zero. Or see what happens when you square the number 101000101.

Of course, the beauty possessed by palindromic squares is only likely to appeal to those who already find numbers a fascinating topic. In order to capture a wider audience, it helps to draw a picture or two—a fact that has not escaped your neighborhood computer vendor. The "demonstration" program that comes with most computer systems usually includes one or more routines to fill the screen with attractive and intricate patterns.

The two accompanying this article have been used as demos for the AT&T DMD 5620 terminal as well as other sophisticated equipment, and late last year AT&T mathematician Peter Maurer made public the simple algorithm that is used to produce such pictures. [*American Mathematical Monthly,* Vol. 94, pp. 631–645.] For obvious reasons, the algorithm is referred to as "The Rose."

Here, in terms that can easily be transcribed to any standard computer language, is what you need to do in order to produce such pictures on your own computer.

The program uses whole number variables N, D, and A, and real number variables X, R, T, old$X$, old$Y$, new$X$, new$Y$.

At the start of the program, set the variables N and D equal to two whole numbers between 1 and 359 (inclusive), and set A, old$X$, and old$Y$ equal to zero.

Now you set up a loop consisting of the following steps.

1. Set A equal to $A+D$. If $A > 360$, replace A by the remainder obtained when you divide A by 360 (that is compute $A \bmod 360$ and set A equal to the result).

2. Calculate $N * A$, then reduce it $\bmod\, 360$ (i.e., take the remainder on dividing $N * A$ by 360), convert the result from degrees to radians, and set X equal to the final result. (To convert from degrees to radians, multiply by 0.01745.)

3. Set R equal to the sine of X.

4. Convert A from degrees to radians, and set T equal to the result.

5. Set new$X$ equal to $R * \sin(T)$ and set new$Y$ equal to $R * \cos(T)$. (This converts the polar coordinates (R, T) to rectangular coordinates (new$X$, new$Y$).

6. Draw a line from the point (old$X$, old$Y$) to (new$X$, new$Y$).

7. If A is equal to zero, then stop, else set old$X$ equal to new$X$, old$Y$ equal to new$Y$, and go back to step one.

The only input the program requires are values for the two integer variables N and D. Different choices of these two numbers produce often strikingly different patterns. For some choices the pattern turns out to be fairly rudimentary, sometimes just a single dot. For other values you will obtain pictures every bit as attractive as the two shown here. For the record, these are produced by taking $N = 4$, $D = 34$ for the first one and $N = 5$, $D = 97$ for the second.

They may not smell as sweet, but the roses you produce on your computer screen can bring every bit as much pleasure as the real thing.

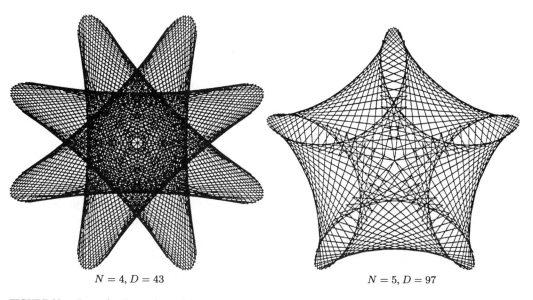

$N = 4, D = 43$ $N = 5, D = 97$

FIGURE 33.   Roses for the mathematician.

111

Theories that all fall down

June 16, 1988

Spend an evening playing around with the familiar whole numbers and there is no telling what you might discover. For all their seeming simplicity, these most fundamental of all objects from the mathematician's domain offer a wealth of strange and unexpected properties.

Take the case of *Micromaths* reader P. H. Kopper of Eastbourne, who, inspired by the recent articles on Fermat's Last Theorem [see Chapter 108], stumbled across the following curious fact.

If you start adding together the cubes of the positive whole numbers, the result in each case is the square of the sum of those numbers. For instance: $1^3 + 2^3 = 9 = 3^2$ (and $3 = 1 + 2$); and $1^3 + 2^3 + 3^3 = 36 = 6^2$ ($6 = 1 + 2 + 3$). Moreover, this pattern seems to continue indefinitely. Adding together the numbers 1 to 4 gives 10, and adding together cubes of the numbers 1 to 4 gives the answer 10^2 Add the numbers from 1 to 5 and you get 15, while adding the cubes from 1 to 5 gives 15^2. And so on.

This is curious, but is it anything more than a mere accident? Does the same pattern really continue forever or, when the numbers get large enough, do sums of successive cubes no longer turn out to be equal to the square of the straight sum of those numbers?

The kind of evidence you can muster in half an hour, sitting in front of the fire with a pencil and paper adding together cubes, might be highly suggestive, but this kind of approach can never provide a cast iron proof. Nor will it help you very much to use your home computer. You may be able to extend your calculations from the cubes of the first dozen or so numbers up to cubes of the first few hundred, but with an infinitude of numbers to deal with, no such piecemeal approach can hope to succeed.

Faced with the task of establishing some fact to be true of all the positive whole numbers, a case by case analysis cannot by itself be adequate. What is required is some rigorous, mathematical reasoning. Though the professional mathematician has a small arsenal of techniques that can be brought to bear on problems of this kind, by far and away the most powerful and widely applicable is the so-called Method of Induction.

Imagine a row of upright dominoes, placed so that any one that topples will knock over its neighbor. They will all topple over once the first one falls.

Mathematical induction works in the same way, only in the case of whole numbers, the "chain of dominoes" is infinitely long.

You want to prove that something or other is true for all whole numbers. Suppose you can show that the property concerned is true for the first one or two whole numbers, and can prove that if the property holds for any number n then it automatically holds for the next number $(n+1)$. Then it follows that the property does in fact hold for all numbers.

Using the method of induction, it is possible to prove that for any whole number, say 356, if you were to add together the cubes of all the numbers from 1 to 356, the answer would be the same as you would get from adding together all the numbers from 1 to 356 and then squaring the result. So the *Micromaths* reader from Eastbourne has indeed discovered a genuine mathematical truth about whole numbers.

Or perhaps I should say rediscovered, since this particular result has been known for many years. But as in many other walks of life, while there is an extra kick to be gained from being the first, the pleasure of discovery is as great for the 101st person as it is for the second and third.

Finally, the recent spate of advances in the factorization game continues unabated. Hot on the heels of record-breaking computer factorizations of large numbers by Silverman in the US and then te Riele's team in Amsterdam [Chapters 107 and 109], Richard Brent

of the Australian National University in Canberra, has just announced the discovery of two new prime factors of the astronomically large Fermat number $2^{2048} + 1$. This number is the eleventh in a list described by the 17th Century mathematician Fermat, who thought erroneously that all the numbers on this (infinite) list were prime. Only the first five are known to be prime. The next 15 on the list are all known not to be prime, but only the first four of those have been completely factored into primes.

There is still some remaining doubt. What Brent has done is split the eleventh Fermat number into a product of five smaller numbers, two with six digits (these were already known), one with 21 digits, one with 22 digits, and a final one with 564 digits. Of these, the first four are known to be prime, but the final, big number has not yet been proved to be prime.

Postscript.   See Chapter 139 for an update on the factorization of Fermat numbers.

112

Can a smart computer ski?

July 14 and 28, 1988 (merged)

The joke is an old one. Coming home late one night I see my neighbor crawling around the pavement beneath a streetlamp.

"Lost something?" I ask.

The man glances up at me. "Yes, my house keys. I'm locked out."

Helpfully, I start to search the ground alongside him. "Are you sure you lost them around here?" I ask.

My neighbor gives me a surprised look. "Oh no," he explains, "I'm pretty sure I lost them further down the street. I'm only looking here because the light is so much better."

No sensible person would behave like my hypothetical neighbor, at least when it comes to lost house keys. But I am not the first person to see a similarity between this joke and a lot of current work in computer science, in particular the hugely ambitious field known as artificial intelligence.

In 1956, John McCarthy, then a young assistant professor of mathematics at Dartmouth College, New Hampshire, teamed up with his friend Marvin Minsky, of the Massachusetts Institute of Technology, to organize a conference.

McCarthy proposed that: "A two-month one-man study of artificial intelligence be carried out during the summer of 1956 at Dartmouth College. The study is to proceed on the basis of the conjecture that every aspect of learning or any other feature of intelligence

can in principle be so precisely described that a machine can be made to simulate it."

This was not the first time that such ideas had been put forward, though it was the first time anyone had used the term "artificial intelligence." Although not primarily concerned with the actual construction of machines, the nineteenth-century logician George Boole had spent many years trying to formalize the "laws of thought" in a precise, mathematical fashion. He was so successful that the ideas he developed play a key role in the design of computer systems today.

The famous wartime code breaker and computer builder, Alan Turing, also put forward the hypothesis that it would "soon" be possible to program computers that could exhibit intelligent behavior.

But it was that particular phrase "artificial intelligence" in McCarthy's proposal that caught the attention of the rest of the scientific community, to say nothing of the world's press, and thereby established a new field of research. The fundamental idea was that, following the same lines as Boole a century before, mathematical logic could be used as a framework in which formal rules of intelligence could be written down and fed into a computer.

In 1961, McCarthy moved to Stanford University in California, where he helped to set up the Stanford Artificial Intelligence Laboratory (SAIL), and developed a new computer programming language specially tailored for artificial intelligence work: Lisp.

Lisp is now one of the two programming languages most widely used in artificial intelligence, the other being Prolog. (Lisp comes from "list processing," Prolog stands for "programming with logic.")

Among the small group of people who attended the Dartmouth Conference in 1956 were Allen Newell and Herbert Simon of Carnegie-Mellon University in Pittsburgh, Pennsylvania. It is this pair who wrote what is generally regarded as the first ever AI program, the "General Problem Solver."

The fundamental ideas behind this program were to set the guidelines for a whole generation of attempts to develop ever more sophisticated artificial intelligence systems. The general idea can be illustrated by a simple example.

You are writing a computer program to play chess. First you have to figure out a way for the computer to represent the layout of the chess board and the positions of the various pieces on it at any stage of the game. This much is easy.

Then you have to tell the computer what is the aim of the game and what are the admissible moves for each of the pieces. Again, this is easily done, perhaps using a simple language adapted from mathematical logic.

The tricky part is: given the configuration on the board at any stage in the game, how does the program (or the human player) decide which is the best move to make next?

Newell and Simon tackled this problem in the following way. At any stage there is a whole array of possible moves, some of them terrible (in that they will lead rapidly to the game being lost), some not so bad, and maybe one or two very good ones. To discover the best one, look at each possible move in turn and see what would be the result of making it.

Of course, there is still quite a bit of work to be done to make this approach succeed. For one thing, you have to equip the computer with a means of evaluating possible moves (e.g., by material loss or gain) so that it can figure out which one is "best." But techniques developed by Newell and Simon and subsequent workers allow this to be accomplished.

The real problem is the sheer number of possible next moves. Human chess players clearly do not operate by looking at all possible moves. Somehow they are able to narrow down the search to just a handful of "probably good" moves. This, it seems, is where the "intelligence" comes in. And by and large it is true to say that no one has yet been able to figure out how to get computers to perform this step.

Certainly, the successful present-day computer chess programs rely mostly upon raw computing power that enables a search and evaluation to be made of all possible moves.

In terms of capturing intelligence, then, the Newell and Simon approach has hitherto been a failure. But that does not mean the enterprise has been a waste of time. On the contrary, when combined with massive computing power, the Newell and Simon "problem solving" technique of search and evaluation has led to the development of some extremely useful computer systems that can help make medical diagnoses, design electronic circuits, handle restricted forms of everyday language, and even prove simple mathematical theorems. A more recent development due to Newell and Simon is a problem solving system known as *Soar.*

In the four years or so since its inception, Soar has shown itself to be both versatile and powerful at performing the kinds of task that other existing AI programs can do. Of that there seems little doubt.

Far less clear-cut is the claim made by Newell in an article in the magazine *Science* last month, that programs such as Soar bring us significantly closer to the original AI goals of a workable scientific theory of intelligence and cognition, and thus genuinely intelligent computer systems.

For despite the success of programs like Soar at certain kinds of limited tasks, they all take the approach that intelligence consists of (or at least can be mimicked by) the application of a collection of rules, in order to choose the best among an enumerated range of possibilities.

In certain situations, this approach works. But as far as AI goals are concerned, it seems that this is just another case of the man looking for his keys in the wrong place because the light is better there.

The question is, just how much of human cognitive behavior can be handled using rules? The answer may be: not much.

To give just one simple example, when you learn to ski, you start off by learning, and trying to follow, a whole series of rules: bend your knees, keep your weight forward, edge your skis downhill, and so on. The result is an awkward action that looks just what it is: the mechanical following of rules. Good skiing comes later, and consists not of your having mastered the rules better, or having acquired more rules, but rather of your having learned to do without the rules.

That is an opinion, not a fact. But if correct, and if a great deal of skillful cognitive activity is not rule-based but the very opposite, then the Newell and Simon problem solving approach to artificial intelligence will never achieve success outside a limited domain, and any possibility of simulating genuine cognitive behavior will require quite new methods that provide illumination further down the street.

113

Why odd cannot be perfect

August 11, 1988

There do not seem to be many perfect numbers. In fact, to date only 31 have been discovered. Any new ones will have to have many thousands of digits, and their discovery will only be possible with the aid of extremely powerful computers. Still, such discoveries are possible, with the last one made only a few months ago [Chapter 105].

But the discovery of an odd perfect number would seem to require more than a new supercomputer. It is now clear it would take a miracle.

The story begins, as do so many stories involving the positive whole numbers, in Ancient Greece. The Pythagoreans (around 500 B.C.) called a number "perfect" if it is equal to the sum of all its exact divisors except for the number itself.

For example, the number 6 is perfect, since the exact divisors of 6 (other that 6) are 1, 2, 3, and the sum of these three divisors is 6. The next perfect number is 28. The divisors of 28 are 1, 2, 4, 7, 14, and their sum is 28.

There has been no lack of interest in this curious phenomenon among the mathematically minded, and the search for perfect numbers has continued throughout the ages. The third and fourth perfect numbers were discovered some time before A.D. 100. These are 496 and 8,128. The next one was not found until the 15th century, the 8-digit number 33,550,336.

One fact that was observed quite early on in the hunt was that all the perfect numbers were even numbers, though from the definition, there is no reason to suppose this must always be the case.

Indeed, all the perfect numbers discovered end either with a 6 or with 28 (the first two perfect numbers), and it is nowadays known that any even perfect number that might be discovered at any time in the future will also end in either a 6 or in 28.

In fact, there is quite a lot known about even perfect numbers. For instance, every even perfect number is equal to a sum of the form $1 + 2 + 3 + 4 + \cdots + n$, for some number n. For example, $6 = 1 + 2 + 3$ and $28 = 1 + 2 + 3 + 4 + 5 + 6 + 7$. And again, every even perfect number beyond 6 is equal to a sum of consecutive odd cubes; that is, a sum of the form $1^3 + 3^3 + 5^3 + \cdots + n^3$, for some odd number n. For example, $28 = 1^3 + 3^3$ and $496^2 = 1^3 + 3^3 + 5^3 + 7^3$.

One more: for any even perfect number n other than 6, if you add together all the digits in n, then add together all the digits in that number, and so on, you will eventually end up with the number 1.

Much of what is known about even perfect numbers comes from a fairly deep discovery made by the 18th-century Swiss mathematical genius Leonhard Euler. The ancient Greek mathematician Euclid had demonstrated that if you start with some number k and calculate the number $2^k - 1$, and if that new number turns out to be prime, then the result of doubling that new number exactly $k - 1$ times will be a perfect number. Euler proved that every even perfect number can be obtained in this seemingly bizarre way.

Indeed, this has led to the discovery of all the recent perfect numbers. The search reduces to the hunt for the prime numbers of the form $2^k - 1$. Whenever such a prime number is found, the appropriate number of doublings produces a perfect number straight away.

But what of odd perfect numbers? No one—neither Euclid, nor Euler, nor anyone since—has been able to prove conclusively that perfect numbers have to be even. It remains a possibility that there are odd perfect numbers. But a recent computational assault on this problem suggests that even if there are odd perfect numbers we may never be able to calculate one.

In fact, this was already starting to seem likely back in 1957, when a chap called Kanold carried out a huge calculation to show that there are no odd perfect numbers having fewer than 20 digits. In 1973, Tuckerman pushed this limit up to 36 digits, and then Hagis went to 50 digits.

In 1976, it was announced that further work had extended the bound to 200 digits, but the claimants were never able to produce convincing evidence for this claim, and so it was widely discounted.

But as a result of the recent work mentioned earlier, it is now known that there can be no odd perfect numbers with fewer than 160 digits and very likely none with less than 240 digits. Running a Pascal program on a DEC VAX 11/750 computer in Australia, Richard Brent and Graeme Cohen have obtained the bound of 160 digits. With some additional help from factoring expert Herman te Riele of the Netherlands, they are currently nearing the end of a long computation that will produce a bound of 240 digits.

So if anyone comes along and claims to have discovered an odd perfect number, the first thing to do is remember the old market traders' motto: never mind the quality, feel the width. If the number has fewer than 240 digits, it certainly ain't perfect.

114

Probability rod for your back

August 25, 1988

Earlier this year I managed to inflame passions among *Micromaths* readers by posing an innocuous looking probability question. (The coins in the drawer problem [Chapter 103].) Here is another.

While taking down our lounge curtains for cleaning the other day I accidentally dropped the rod (one of those trendy glass affairs) causing it to break in two places. What is the probability that these three pieces could be put together to form a triangle?

Happy with that one? I hope so, because here comes the sting in the tail. Taking down the second curtain rod I dropped this one as well, and it broke, this time in just one place. I was so annoyed at having broken a second rod that I picked up one of the two pieces at random and threw it down hard, causing it to break into two. Now what is the probability that the three pieces can be put together to form a triangle?

At first glance you might think that the second problem boils down to the same as the first. Everything is random, and the result in each case is that the curtain rod ends up broken into three pieces. But things are not as they seem, and in fact the two problems have quite different answers. At least, that is what I am claiming. Perhaps you think differently. No doubt you will be eager to let me know. I'll give my solution in a month or so's time.

I should add that in both the problems you should assume that each break in the glass rod occurs at a completely random position along its length. Doubtless the laws of physics prevent this from really happening but the poor problem solver does need something to go on.

And, in that last remark, you have the glimmerings of a danger that lurks behind every attempt to solve a real-life problem using mathematics. The danger is all the worse when a computer is used, as it gives a reassuringly high-tech, neatly printed gloss to the whole affair. That danger is that in the inevitable, initial process of turning the problem from a real-life one to a mathematical one, something crucial is lost.

The following example, taken from an old puzzle book, is a case in point. Two hikers are camping (some distance apart) quite near to the famous Straight River, so called because it exhibits not a single bend in all of its 10-mile length. One of them, Mike, looks out of his tent one morning to see that the tent of his friend, George, is on fire. Since George is not around Mike grabs a bucket and heads for the river for water to douse the fire before it gets out of hand. The question is, towards which point of the river should Mike go to minimize the time it takes him to get there and back with the water?

The traditional answer to this teaser is that Mike has to minimize the total distance he has to cover. Using some simple geometry you will see that this is achieved by his imagining George's tent reflected through the line of the river to a point on the opposite side, and heading directly towards that imaginary location.

True enough, the geometry works out fine. This solution does indeed minimize the distance Mike has to run. The mathematics is fine. But the answer is clearly wrong in terms of the original question.

What you are supposed to do is figure out the quickest way to get a bucket of water from the river and back to George's tent, starting from Mike's tent. There is no mention of shortest distance. The writer of that particular puzzle (which was printed along with the answer just given) slipped happily, and obliviously, from shortest time to shortest distance.

Since heading for the river involves a fresh Mike carrying an empty bucket, and coming back involves a tired Mike (after that dash to the river) carrying a full bucket, there is no way that minimizing the distance can minimize the time.

The best strategy, surely, is for Mike to head as fast as possible for that point on the river bank that is closest to George's flaming tent: in other words, minimize the distance the water has to be carried.

It is not a mathematical solution, but in terms of saving poor George's tent, by far the most sensible.

115

Great lengths and hidden powers
October 6, 1988

Imagine you are lecturing on a difficult math problem. You have tried to solve the problem yourself, so you know how hard it is. So you promise to pay $1,000 to anyone who can solve it.

Your money seems to be safe. But sitting in your audience is a tenacious individual who is prepared to go to great lengths to find a solution, including enlisting the help of a Cray supercomputer. Within two weeks, your listener returns ...

Not only that but, as the man gently points out, it was not $1,000 that you offered but $10,000—a slip of the tongue that was recorded for posterity on videotape.

This happened to the British mathematician John Horton Conway a few weeks ago.

Dr. Conway, who moved from Cambridge to Princeton University in New Jersey in 1986, is well known to computer users as the inventor of the computer game "Life."

The problem he posed to the 500 or so professional mathematicians and scientists attending his lecture at a meeting held at AT&T's Bell Laboratories in Murray Hill, New Jersey, on July 15, was a typical Conway teaser, closely related to an old problem of Fibonacci.

In 1202, the Italian mathematician Fibonacci of Pisa introduced what is nowadays known as the Fibonacci sequence. This sequence starts off with two 1s, and there after grows according to the rule that each new number is the sum of its two predecessors. Thus, the first few numbers in this sequence are: 1, 1, 2, 3, 5, 8, 13, 21, 34, 55.

Conway's problem concerned an amended version of the Fibonacci sequence, obtained as follows. Start off with two 1s, but now the rule for calculating each new number is: take the last number you have, call it n, and then add together the nth number in the sequence and the nth number from the end.

This gives you a sequence that starts off like this: 1, 1, 2, 2, 3, 4, 4, 4, 5, 6, 7. To find the next number count forward 7 places to the number 4, count back 7 places to the number 3, and add 4 and 3 to get 7. And so on.

The first thing you will discover if you work out a few more terms (it's easy to program on a micro), is that each new number in the sequence is never greater than the previous one by more than 1.

A bit harder to spot, but just as easy to check (by computer) is that if you divide each number in the sequence by its position in the sequence the answer you get is always close to one half. This turns out to be no mere accident. Dr. Conway managed to prove conclusively that the further along the sequence you go, this ratio gets steadily closer to one half.

The question Dr. Conway asked his audience at Bell Labs was this: from which point in the sequence is this ratio always within 0.05 of one half?

Before you rush to your micro to find out, I should point out that that figure of 0.05 is a fairly tight bound, and at the time of his lecture, Dr. Conway was convinced there was no way of getting the answer. Hence the $10,000 wager.

But Conway reckoned without Dr. Colin Mallows, another expatriate Brit, now work-ing at Bell Labs. Bringing to bear all the modern techniques of statistical data analysis, pattern recognition, computer generated graphs, and a supercomputer, Mallows had the solution within two weeks. The critical position in the sequence is number 3,173,375,556. So Conway took a deep breath and wrote out a check for $10,000. But Mallows declared himself happy to accept the intended $1,000 and to keep the larger check as a souvenir.

116

The anguish in the broken curtain rod

October 20, 1988

As expected, the probability teaser I posed on August 25 [Chapter 114] caused many readers a great deal of anguish. But now, at last, relief is at hand. Or is it?

The problem was this: while taking down a glass curtain rod, I dropped it, causing it to break in two random places. First question: what is the probability that the three pieces can be put together to form a triangle?

Then, while taking down a second curtain rod, I drop that one as well, and this time the rod breaks randomly into two pieces. This annoys me so much that I pick up one of the two pieces at random and throw it to the floor, causing it to also break randomly into two pieces. Second question: what is the probability that the pieces can form a triangle this time?

The sting is that the two answers are not the same—or so I claim, though there was no shortage of readers to disagree with me. After all, their argument ran, the net effect in either case is that the rod breaks into three pieces in a totally random fashion, so how can the answers be different?

A more subtle variant suggested by two readers is: what if in the first case, when "nature" does all the breaking (unaided by an annoyed Keith Devlin), the breaks occur shortly after another, with "nature" making the random choice of the piece that sustains the second break? Surely then, the two answers must

be the same? But is this really the same as either of the two original questions?

Here are my solutions. I think they are right. You can judge for yourself. But be warned, there is some good old-fashioned high school mathematics coming up.

To avoid worrying about actual lengths of pieces of the rod, assume the rod has length one unit.

In the first case, let x be the distance of the first break, y that of the second, both measured along the rod starting from one end. Thus the three pieces have lengths x, $y - x$, and $1 - y$ (see below). According to the question, the pair x, y is random among the collection of all such pairs with $0 < x < y < 1$. On the graph shown below, this means that the point x, y must lie in the triangular region indicated. To get a triangle, the sum of the lengths of any two pieces must exceed the length of the third piece. This gives the three inequalities

$$x + (y - x) > (1 - y)$$
$$(y - x) + (1 - y) > x$$
$$x + (1 - y) > (y - x)$$

Solving these gives the inequalities $y > \frac{1}{2}$, $x < \frac{1}{2}$, $y < x + \frac{1}{2}$. On the graph the points x, y that satisfy these three requirements lie in the shaded region. This clearly has $\frac{1}{4}$ the area of the entire region of possible points x,

y. So the answer to the first question is that the probability of getting a triangle is $\frac{1}{4}$ (or 0.25).

Now for the second question. Here we must be careful, for despite the superficial similarity between the two questions, the randomness assumption in this question is different from that in the first question.

Let x be the length of the piece chosen, at random, after the first break. Let y be the distance along this piece where the second break occurs. The probability of getting a triangle if x is less than $\frac{1}{2}$ is obviously zero. So the overall probability of getting a triangle on the second break can be evaluated on the assumption that x is greater than $\frac{1}{2}$. There are therefore just two inequalities to be satisfied this time

around. They are

$$y + (1 - x) > (x - y)$$
$$(x - y) + (1 - x) > y$$

Solving these gives the inequality

$$x - \tfrac{1}{2} < y < \tfrac{1}{2}$$

In particular, this tells us that y must lie on an interval of length $1 - x$. (Subtract $x - \frac{1}{2}$ from $\frac{1}{2}$). Since the second break is at a random point on the piece of length x, the probability of getting a triangle is given by dividing the length of the interval in which y gives a triangle (just calculated) by the length of the piece, namely $(1 - x)/x$.

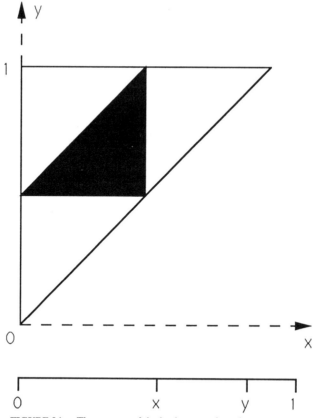

FIGURE 34.   The mystery of the broken curtain rod.

The above expression gives the probability of a triangle for any value of x between $\frac{1}{2}$ and 1. To get the overall probability, you have to add together all these answers for all the x between $\frac{1}{2}$ and 1. Since there are infinitely many such x, this amounts to evaluating an integral: you have to integrate the expression $(1-x)/x$ between the limits $x = \frac{1}{2}$ and $x = 1$. The answer is $\log_e 2 - \frac{1}{2}$, which, according to my calculator, is 0.193.

There you have it. It is not at all the same answer as in the first case, the questions were really quite different from a mathematical point of view. Weren't they?

117

Better by degrees
November 17, 1988

Everyone knows how wonderful today's computers are. Their speed and power increases almost daily, and just around the corner lies an age when these incredible machines will solve all mankind's problems, and lead us to a life of ease and comfort. Or so, at least, one might believe from reading newspaper articles and proposals for research funding!

To quote just one publication, Jastrow's "The Thinking Computer," written in 1982:

"In five or six years—by 1988 or thereabouts—portable, quasi-human brains, made of silicon or gallium arsenide, will be commonplace.

They will be an intelligent race, working as partners with the human race. We will carry these small creatures around with us everywhere ... . They will be Artoo-Deetoos without wheels: brilliant but nice personalities, never sarcastic, always giving you a straight answer—little electronic friends that can solve all your problems."

Well, 1988 is almost at an end and the reality is far from the hype. To date no one has built a single computer system that has performed a single act that could be properly described as "intelligent." They have not even come close. What computers can do, and do well, is act dumb very fast. But even when it comes to elementary arithmetic, the much-vaunted micro can fail where a little human ingenuity can succeed brilliantly.

This week's problem selection is designed to illustrate both the power and the weakness of the computer in the domain in which it can claim to shine most—whole-number arithmetic.

First, here are two simple problems about prime numbers that should cause not a hiccup in your favorite silicon box.

What is special about the prime number 9,431? If you take all the digits in this number, square them, and add the resulting numbers together, the answer is itself a prime.

That is, take the digits 9, 4, 3, and 1, square them to give the numbers 81, 16, 9, and 1, and add together these four numbers to give 107. Like the original number 9,431, this number 107 is also a prime (that is, it can only be divided by itself and 1).

How common are such numbers? Not all that common, it turns out. Of the 1,229 prime numbers smaller than 10,000, there are precisely 237 primes such that the sum of the squares of their digits is again a prime. Use your micro to find all 237 of these four-digit primes. Then see what is so special about the prime numbers 3, 5, 23, 29, and 71.

When you have sorted that one out see if you can find a whole number having distinct digits, all of whose prime factors are single-digit primes whose square has 8 distinct digits.

So far everything can be handled with ease by the humblest of home micros. Now for the sting in the tail. Here is a seemingly simple problem that, if you are not careful, could turn your computer into a night-storage heater.

The formula for converting Fahrenheit to centigrade looks like this:

$$C = \frac{5}{9} \times (F - 32).$$

If you take the Fahrenheit temperature 527°F, it converts to the centigrade temperature 275°C. This is just the original Fahrenheit figure with the first digit moved to the end. But what is the next Fahrenheit temperature with this property that the centigrade equivalent is the Fahrenheit figure with the first digit moved to the end?

The question asks for whole number temperatures only, and they should be positive, though if you insist on trying a computer attack unaided by the human mind, you could take pity on your poor machine and let it have the satisfaction of finding a negative (and physically impossible) solution.

118

The lofty goals of a new mathematics program

December 1, 1988

By the time I arrived at 7:30 P.M., Stanford University's Terman auditorium was full to overflowing. Late arrivals such as myself were advised to move to the two adjacent rooms where the proceedings would be relayed by closed-circuit television. I chose to remain where I was, in the crush standing in the aisles. I wanted a first-hand view of events.

So what was this occasion that had managed to fill such a large lecture hall? What had persuaded so many to abandon their work for the evening and pack themselves in like sardines? (At Stanford, as in the rest of Silicon Valley, evenings are usually for working—along with mornings, afternoons, and in many cases nights as well.)

The answer was a computer program; but not just any computer program. This was a computer program that everyone had been hearing about and was just now becoming available: *Mathematica.*

According to a lot of the word-of-mouth hype that had preceded its appearance, *Mathematica* was a program that was able to "do" all of mathematics. Needless to say, even if you interpret that to mean all mathematical calculations that arise in physics and engineering, which is more or less what the program deals with, it does not achieve that impossibly lofty goal. But it is impressive, and seems destined to become the mathematical package for the professional.

Mathematica is not the first general purpose mathematical package. For example, *Macsyma* is perhaps the best known of a number of programs aimed at roughly the same market. But *Mathematica* has two advantages over its rivals.

The first is that, by coming on to the scene at a later stage, it has been possible to develop it as part of a "grand scheme," taking note of all the predecessors rather than building up the system one step at a time.

The second advantage is the character of its creator. At 28, expatriate Englishman Stephen Wolfram has already discarded his "famous young physicist" label in favor of his new "becoming-famous young software entrepreneur" one.

A faculty member at the University of Illinois, Wolfram conceived of the idea of *Mathematica,* then set up his own company to develop, produce, and eventually sell his brainchild. And it is in the selling that Wolfram has made the major breakthrough that has already made his package the most talked-about of its kind—and will probably make it by far the best selling one.

Partly it was his name. People knew of his precocious talent as a physicist, so there was already considerable interest in the academic world when word got around that he had "gone into computers."

There was also his readiness to go out and peddle his wares. It was Wolfram himself who gave the demonstration at Stanford; his short, dishevelled, bearded appearance and quick-fire, erudite delivery being so obviously that of a passionate researcher that the entire auditorium accepted him as "one of them," not some slick salesman to be wary of.

But most of all, it was getting top hardware companies to sell the package for him. Currently available for Macintosh and Sun microcomputers, *Mathematica* is one of the packages included with former Apple founder Stephen Jobs' sleek new NeXT Computer. True to the form of both Jobs and Wolfram, a NeXT system was in the Stanford lecture hall, ready for the program's creator to put it through its paces.

So what does *Mathematica* do? The answer is practically every kind of calculation. Whole-number arithmetic involving numbers of any size you care to input, floating-point arithmetic of whatever accuracy you desire, algebraic manipulations, matrix arithmetic, solving equations either algebraically or numerically, differentiation, indefinite and definite integration, graph drawing—these and more are all available, and appear to run very efficiently.

I have yet to use *Mathematica* myself, but a colleague has, and the November issue of the *Notices of the American Mathematical Society* carries a number of technical reviews of the program. Apparently the routines really do work well, apart from one or two bugs that will be ironed out in the forthcoming, upgraded version.

Further, there's a "textbook" feature which turns *Mathematica* into an interactive teaching tool of considerable sophistication. This makes Wolfram's package a formidable piece of software for the higher education market, to say nothing of the world of engineering and technology. In short, it appears to do for the whole of manipulative mathematics what the spread-sheet did for business managers.

For that reason, *Mathematica* will probably draw as many critics as it does supporters, but one thing seems clear: it is going to change forever the way people do mathematics.

119

A problem that cuts across conventional boundaries
December 15, 1988

When Stanford University decided to change one of its first year arts courses earlier this year, the associated faculty discussions were reported on the front page of the national newspapers, made the nightly television news, and even drew public comment from the US Secretary for Education.

How is it that a private university with about the same number of students as, say, the University of Leeds, can attract such attention? The answer lies in the track record of the place, and the surrounding Silicon Valley area, when it comes to innovation. Broadly speaking, what Stanford does today, the rest of America will do tomorrow, and much of the world the day after.

Its track record has resulted in this numerically small university having a current annual research income of some $233 million. So people pay attention to any developments that take place on this attractive and extensive campus, nestled between the San Fransisco Bay and the Santa Cruz Mountains, just an hour's drive south of San Fransisco.

In particular, the recent decision to extend the study of "symbolic systems," by introducing a graduate program to supplement the three-year old undergraduate scheme could well turn out to be significant for computing and communications.

Symbolic systems? First, I must declare an interest. As regular readers will know, I am on a two-year visit to Stanford, and both my research and teaching here are connected with the symbolic systems program. Back in England, I had started to do research in symbolic systems before I knew it had this name, or the extent to which the subject had developed at Stanford. So I am biased.

Symbolic systems is a new approach to the study of information, that cuts across the traditional boundaries of computer science, artificial intelligence, linguistics, psychology, neuroscience, philosophy, and logic.

The kinds of issues that are studied include questions such as: What is information? What is mind? Does intelligence require a mind? How are minds related to brains? Is intelligence fundamentally biological, or is it possible to create artificial devices that handle information in an intelligent manner?

Of course, most of these questions are not new. But until recently, the absence of any framework for scientific study left them almost exclusively in the domain of the philosopher. What has changed this is our new-found friend the computer, and our attempts to design machines that think.

The emergence of symbolic systems as a subject in its own right comes from three lines of research that all have their origins in the mid-1950s.

Starting with a now-famous conference held at Dartmouth College in New Hampshire in 1956, the work of Minsky, McCarthy, Newell, and Simon led to the discipline of artificial intelligence. Its aim is the construction of intelligent devices.

At the same time, psychologists were starting to use the computer in order to study and model the human brain, in an effort to understand the phenomenon of cognition. This led to the establishment of cognitive science, a field generally acknowledged to have begun with a 1956 conference at MIT. It was helped in its development by substantial support from the Sloan Foundation during the 1970s.

The work of yet a third group had fundamental links to both artificial intelligence and cognitive science.

Research in linguistics, starting with Chomsky's work in 1956, led eventually to the search for computer systems capable of "understanding" (parsing) and translating natural languages. Thus computers began to be used to assist in the unravelling of the mechanisms by which human language works.

Symbolic systems spans all three areas. It has a great deal in common with each of artificial intelligence, cognitive science, and linguistics, but is not rooted to the operation of computers (as is the first) or to the human mind (as are the second two). Whereas the mind simply provides the device that artificial intelligence seeks to simulate, and the computer simply provides a metaphor for the mind, or a tool to simulate mind, in cognitive science, both the computer and the mind are part of the subject matter in symbolic systems research.

The science of symbolic systems starts with the observation that agents (be they human, animal, or artificial) use symbols to represent information about their environment, and to communicate that information to other agents. The word "symbol" includes not only written symbols, as in the case of written language, but spoken language, communication using very primitive forms of signals such as grunts or cries, computer languages, the distributed representations in neural networks, and the electrochemical symbols manipulated in the brain.

But what exactly is this "information," and how can it be "represented" or "carried" by symbols? These are the questions that immediately arise, and they clearly lie at the very heart of any theory of language, communication, or computation. Unfortunately, a few moments thought should be enough to convince you of the difficult task facing anyone trying to find answers.

After 30 years of effort, it seems unlikely that a solution will be found by someone trained in any one of the existing scientific disciplines. The problems cut across too many traditional boundaries.

Though a significant number of researchers from all the relevant disciplines at Stanford and elsewhere are increasingly starting to pool their talents and look for a solution together, progress is still painfully slow.

The main hope in starting the new graduate program in symbolic systems at Stanford is that this will produce a new generation of scholars who will be able to succeed.

- *Time*'s Man of the Decade:
  Mikhail Gorbachev
- Tiananmen Square
- Tom Cruise: *Born on the Fourth of July*
- Hugo chews up South Carolina
- Earthquake rumbles thru Oakland
  and Frisco
- Exxon Valdez
- Emperor Hirohito dies
- Voyager nears limits of solar
  system
- Berlin Wall falls
- *Batman*
- *Roseanne* hits #1 on
  Nielsen parade
- Deborah Norville vs. Jane
  Pauley
- Pete Rose banished from baseball
- *The Satanic Verses*
- USS Iowa explosion kills 47
- Zsa Zsa arrested for slapping cop
- Bakker guilty
- Leona Helmsley
- Polish elections go to Solidarity
- *Field of Dreams*
- *When Harry Met Sally*
- Ferdinand Marcos dies

1989

120

Greek insights in a prime challenge
January 5, 1989

In 1640, in a letter to a colleague, the great French amateur mathematician Pierre de Fermat claimed to have proved that any number of the form "one more than 2 raised to a power 2^n" is a prime number. In honor of this claim, such numbers are known as *Fermat numbers*. For any whole number n, the nth Fermat number is 2 raised to the power 2^n, plus 1. This number is usually denoted by F_n.

It is easy to check the first few cases of Fermat's claim. The first Fermat number, F_0, is equal to 2 raised to the power 2^0, plus 1. As 2^0 equals 1, this gives $2 + 1$, which is 3, a prime number. The next one, F_1, is 2 raised to the power 2^1, plus 1. As 2^2 equals 4, the answer is 5, which is also a prime.

Things soon start to get a bit trickier, on account of the size of the numbers involved. F_4 is equal to $2^{16} + 1$, which is 65,537. This is also a prime number. But the next Fermat number, F_5, is the 10-digit monster 4,294,967,297. Is this prime or not? According to Fermat's claim it must be.

In fact F_5 is not a prime. As the Swiss mathematician Euler discovered in 1732, the number 641 divides into F_5. Euler made this discovery using the fact (which he himself established) that any number that evenly divides into a Fermat number F_n has to be 1 more than a multiple of 2^{n+2}.

Euler's result has since been used to find (by simple search) factors of various other Fermat numbers, the largest to date being the giant $F_{23,471}$, which is divisible by $(5 \times 2^{23473}) + 1$. This was discovered by a chap named Keller in 1984.

Some idea of the size of numbers involved can be gleaned from the fact that a pile of F_6 tenpenny pieces would stretch from the surface of the Earth to the nearest star, Proxima Centauri, some four light years away.

At any rate, Fermat was wrong in claiming that all Fermat numbers are prime. But just how wrong was he? Well the only Fermat numbers known to be prime are the first five, F_0 to F_4, and nowadays the smart money goes on the conjecture that all remaining Fermat numbers are non-prime.

The numerical evidence for this is pretty small. All that is known is that each of F_5 to F_{21} are non-prime, and that about 60 other Fermat numbers are also non-prime. The smallest number whose primality status remains unknown is F_{22}.

The fact that F_{20} is not prime was proved as recently as 1987, when Jeff Young of Cray Research in Minnesota and Duncan Buell of the Supercomputing Research Center in Lanham, Maryland, ran a Cray-2 supercomputer for 10 days to obtain this result, a calculation involving approximately 2.3×10^{13} floating-points multiplications and divisions.

The method used by Young and Buell did not make use of the Euler trick mentioned a moment ago. They utilized another approach, known as Pepin's Test. This says that to see if a Fermat number F_n is prime or not, calculate the remainder on dividing 3 raised to the power $(F_n - 1)/2$ by the number F_n. (There is a relatively straightforward way of carrying out this computation without involving any numbers greater than F_n.) If the answer is -1, F_n is prime; otherwise, F_n is not prime.

One interesting feature of Pepin's test is that even if it tells you the Fermat number is non-prime, and therefore has a proper factor, it does not tell you what that factor is. Thus F_{20} and F_{14} are the only two of the non-prime Fermat numbers F_5 through F_{21} for which no factor is known.

The credentials of the two workers who attacked F_{20} gives some indication why interest continues in Fermat numbers. They provide excellent problems for testing new hardware and software, and the result can be easily verified by duplicating the calculation on another machine. (This was done in the case of F_{20}.)

Another reason lies in an astonishing result proved by Karl Friedrich Gauss in 1796 (when he was still a 19-year-old student) that linked Fermat numbers with the ancient Greek's construction of geometric figures.

What Gauss showed was that it is possible to construct, using ruler and compasses only, a regular polygon having N sides, if and only if N is either a power of 2 or else a power of 2 multiplied by some collection of distinct Fermat primes.

A particular consequence of this result is that, unbeknown to the ancient Greeks, it is possible to construct a regular polygon having 17 sides. In memory of this one and only advance in ruler and compass construction since ancient times, a monument erected to Gauss at his birthplace in Germany now bears a regular 17-sided polygon. It's one of the hidden links that connect today's high-tech mathematics with that of less frenetic times.

121

The programs of unshakable absolute certainty

January 19, 1989

It used to be so simple. Because of their reliance on measurement and experiment the scientist could never be sure that their theories were totally correct, but the mathematician had no such problem. Resting in the secure foundations of the notion of *proof,* mathematical truth had an unshakeable, absolute certainty to it.

If you wanted to know if some mathematical assertion (say Pythagoras' theorem on right-angled triangles) was true, you had to discover a proof for yourself. Once you had a proof, you knew for certain that the result was true. It was not a question of judgement, of evaluating the evidence. A correct proof was a correct proof, and no two mathematicians would disagree on the matter.

Of course, things were not quite that simple with long proofs, stretching over many pages, and involving very subtle reasoning, it was always possible that there was a mistake in the logic that no one, neither the proof's creator nor any subsequent readers, had spotted. This has happened more often than most mathematicians care to admit. But human error aside, the notion of a mathematical proof stood alone in the sciences as an infallible method of verification. All that changed with the computer. The first significant use of computers in the discovery of a proof occurred in 1976, when Kenneth Appel and Wolfgang Haken of the University of Illinois used a computer to find a proof of the elusive Four Color theorem.

The proof discovered with the aid of the computer was far too long to be checked by human beings. If you wanted to convince yourself that the result was indeed correct, you had to take a look at the program and convince yourself it did what its authors said it did. As anyone who has tried to write an even moderately complicated program will know, this is far from easy. And, even today, over 12 years later, there is still some skepticism voiced about this result. But there is no resisting progress, and with the development of ever more powerful computers, and their increasing availability, more and more mathematicians are using them to assist in solving difficult problems. Two recent announcements emphasize that not only the way mathematics is being done, but the very nature of mathematics itself, has been changed forever.

The first example concerns perfect numbers. A perfect number is a whole number that is equal to the sum of all its exact divisors. For example, 6 is a perfect number, being equal to the sum of 1 and 2 and 3, its exact divisors. So is 28. To date, some 31 perfect numbers have been discovered, and they are all even numbers.

An old question is whether there can be any odd perfect numbers. Though such giants of mathematics as the 18th century Swiss genius Leonhard Euler have attempted to answer it, no one knows.

Richard Brent and Graeme Cohen have not resolved the matter, but have shown that if there are any odd perfect numbers, they must be extremely large indeed. In fact, any such number will have to require at least 160 digits to write out in full.

Or so, at least, the mathematical world will have to accept from the computer's mouth. For as Brent and Cohen say in their announcement of the result, "The proof is almost entirely computer generated . . . ." The proof is far too long for any human to check all the steps.

The same situation arises with another recent result. Just before Christmas, a team of mathematicians at Concordia University in Montreal announced their solution of a problem that had been open for over two centuries. The problem was whether or not there could be a finite projective plane of order 10. Don't worry if you don't know what this means; neither do most professional mathematicians. It is a bit like a magic square: a table of numbers with various special properties. As with the perfect number result, the use of a computer was essential in finding the solution. (In this case, it was a Cray-1S supercomputer at the Institute for Defense Analysis at Princeton, New Jersey.)

Using computers to perform calculations is one thing, but when they are used to obtain proofs, that changes the nature of mathematics. Should the result of such a computation really be called a proof?

Whether it should or not, such results are being *called* proofs. And increasingly so.

122

Dantzig dimension

February 2, 1989

A powerful new commercial computer system just coming onto the market might be just what you need to help plan your next year's summer holiday. Though the $9 million price tag on AT&T's KORBX computer system might put one or two of you off, it really will help you pick out the best deal on air travel.

Imagine you are planning to visit Australia, or maybe take a holiday in Florida. There you are, standing at the counter in your local travel agent, trying to make sense of the various bargain fares the different airlines offer. It's a hard enough task on its own, made more complicated by the various mileage bonus schemes available on the major carriers.

Your interest is, presumably, to make the trip for the least possible cost, subject to whatever constraints are imposed by your schedule. So you have to juggle dates of booking, days of travel, type of ticket, and who knows what else in order to come up with the best deal.

As anyone who has ever tried to make such a decision will know, the number of possible combinations is usually so great as to make an accurate solution impossible without the aid of a computer. For example, if there are 10 factors to take into account, with each factor giving just three possible choices, the total number of possible combinations of choices is a daunting 59,049.

What you have is a classic case of a linear programming problem. There is some quantity to be maximized (for example, profit) or minimized (for example, cost), where the quantity required to be optimized depends upon a range of parameters, each of which is subject to one or more constraints that limit its possible values.

The "linear" in the name of the problem means that the various dependencies and constraints are all of the simplest mathematical type. But even then you have to be very cunning in the way you set about trying to solve the problem, and most real-life examples that involve more than a handful of parameters, require phenomenal computing power for their solution.

The obvious means of finding a solution is by trying all possible values of the different parameters (or at least a good representative sample of these), and seeing which combination gives the optimal result. But the number of possible combinations makes this impractical for all but the simplest cases. You need some clever mathematics to get you by.

Fortunately, there is some. The best-known mathematical method, and until now by far the most widely used, is the Simplex Method, invented by the American mathematician George Dantzig back in 1947.

Dantzig's idea was to convert the optimization problem to a geometric one that involved finding a certain very special corner of an n-dimensional polytope. A polytope is just a higher dimensional generalization of familiar two-dimensional, straight-edged geometric figures such as triangles, quadrilaterals, pentagons, and so on, and the three-dimensional, straight-edged geometric solids such as the tetrahedron, the cube, and so on.

Though the three-dimensional universe we live in forces us to stop short of the mind boggling idea of a four-dimensional analog of, say, a tetrahedron, the mathematician faces no such difficulty. Polytopes of any number of dimensions can be dealt with quite happily using modern mathematics. Moreover, it is easy to train a computer to handle them.

So following Dantzig, you transfer your original problem to a polytope. Then you are faced with the new problem of finding that special corner. Unfortunately you don't know where this particular corner is until you find it. It is just like a game of hide and seek.

Dantzig's Simplex Method finds its way to the special corner by starting from any old corner, then using a simple route-finding device that allows you to work your way along the edges of the polytope until you find it.

Of course, the number of possible routes is astronomically large for all but the simplest of cases. In principle this huge number of possibilities could overwhelm the Simplex Algorithm, but in practice this almost never occurs.

But mathematicians have always been on the look out for a better method. In particular, organizations such as the giant AT&T communications company, faced with such tasks as finding the most cost-effective way to connect together millions of telephone subscribers spread all across North America, are very interested.

Four years ago, this interest paid off when the 28-year-old AT&T mathematician Narendra Karmarkar discovered a brand new

(and somewhat bizarre) method of finding Dantzig's special corner.

Following the discovery, there was a period of intense disagreement as to just how efficient the new method was compared to the familiar Simplex Method. AT&T pressed ahead with an extensive testing and development program, aiming to use it as the powerhouse of a brand new computer system designed for sale to the major airlines, oil companies, and other large corporations for whom linear programming represents their very life blood.

In May, AT&T received three patents covering the computer implementation of the Karmarkar Algorithm, and in August announced that they were ready to market their new system. What your $9 million buys, the company claims, is a significantly more powerful tool for solving large scale optimization problems.

Karmarkar's method avoids the possibly lengthy task of meandering along the edges of the polytope step by step in search of the special corner, by leaving the outer shell altogether and simply heading into the interior. In other words, it takes a short-cut.

The snag is: since you don't know where the special corner is, how do you know which way to go once you've left the edge?

The Simplex Method might involve a long trip, but at least there is a device to help you decide which way to turn whenever you come to a crossroad. The Karmarkar Method takes you straight into the emptiness of n-dimensional space. The trick is to make use of a highly ingenious mathematical "direction finder." Discovering how to construct such a direction finder was the critical advance made by Dr. Karmarkar in his 1984 breakthrough.

As a guide through n-dimensional space to an unknown target, that $9 million price tag suddenly doesn't seem too high.

123

The private truths

March 23, 1989

One of my favorite cartoons shows a picture of Sir Isaac Newton, sitting beneath the apple tree in his East Anglian garden, his hand gently rubbing the top of his head as a large apple bounces off it. The caption reads: "Gravity: You may not like it, but it's the law."

It was, of course, Newton who discovered the law of gravity, so the legend goes, observing an apple fall from a tree in his garden. It was a discovery that was to change the course of history, marking a significant advance in man's understanding of the world. But in what sense is this law Newton's? He surely cannot own it. Nor can he be said to have invented it.

When we speak of Newton's Law, we are simply acknowledging his recognition, or discovery, of this universal fact of nature. It follows that if Newton were alive today, and were employed by, say, IBM, neither he nor his employers would be able to claim ownership of, or file patent for, his great discovery. Fundamental laws of nature are public property, available to all, both before and after their discovery. That, too, is the law—only this time not a law of nature but the man-made law of copyright and patent.

But when talk switches from laws of nature to laws of man, you can be sure that the lawyers are in for a field day.

"Algorithms" is the buzzword in parts of the US legal profession, following some patent awards that have stirred up controversy in both commercial and academic research circles.

An algorithm is the scientific name given to any set of instructions for performing some operation. For example, a step-by-step recipe for baking a cake, a step-by-step instruction manual for servicing a car or erecting a garden shed, or the instructions for performing some particular calculation. A computer program written to perform some particular calculation is also an implementation of some algorithm. Once you have worked out the algorithm to solve your problem, writing the program to implement it is largely routine.

Not surprisingly, given the amount of money at stake in today's computerized world, a clever new algorithm can be worth millions of dollars to the company that discovers it. It is thus an asset the company will want to protect by whatever means are available—such as the paws of patent. But what exactly are you patenting when you file patent for an algorithm? Indeed, is an algorithm something that can be patented?

Typically, the specification of an algorithm involves the writing down of various mathematical equations that express some universal facts about the world. But doesn't that begin to sound like Newton's law of gravitation, something that is surely not patentable?

Indeed it does, and as early as 1939, the US Supreme Court ruled that "Scientific truth, or the mathematical expression of it, is not a patentable invention."

But times change. Handing down its ruling in the 1981 case of "Diamond versus Diehr," the US Supreme Court eased its hard line on mathematical expressions by observing that the mere presence of an algorithm should not of itself render an invention unpatentable. "It is now a commonplace that the application of a law of nature or mathematical formula to a known structure or process may well be deserving of patent protection," the court ruled. Since then a number of patents have been granted for what are, to all intents and purposes, straight algorithms, or mathematical equations.

This year patents have been awarded to Pierre Duhamel for his Discrete Cosine Transform (useful in the compact storage of video data) and to Eastman Kodak for a Picture Compression Algorithm that allows pictures to be stored compactly in computers with a minimal amount of distortion.

These patents represent a potentially damaging change in the environment in which mathematicians and computer scientists carry out their research. No research is done in a vacuum. Each new discovery builds on the work that has gone before, and the invention of new algorithms depends critically on the results of others being freely available to one and all. Should a research group ever feel compelled to keep the results of its work secret, we will have embarked on a path that can only hinder future developments.

So far, this has not happened. But these are early days and no one knows how things will develop in future. The only certainty is that the legal arguments, like the algorithms themselves, will get ever more complicated.

124

A series that hits the buffers
April 13, 1989

Shortly before Christmas, I was sitting round a table in a house in Cambridge, Massachusetts, having dinner with a group of professors from the nearby Massachusetts Institute of Technology.

The party consisted of enough raw brain power to dispose readily of any simple problem that could be explained on the back of a napkin, you would think.

Not so. An innocuous little teaser that the logician George Boolos had come across had everyone baffled, even when he told us the answer. I for one was unable to believe it, and I left Cambridge the next morning to fly back to Stanford still convinced that George was having us all on. I only changed my mind when he sent me a formal mathematical proof of the answer he had claimed. Having seen the proof, I now know the answer has to be correct. But I still don't really believe it. And I don't think you will either.

The problem is ridiculously easy to explain. All it involves is picking fractions (or arbitrary real numbers, for those who know what that means) that lie between 0 and 1, in such a way as to satisfy a certain very simple condition.

First of all, notice that it is easy to pick two fractions, call them a and b, so that a is less than half and b is bigger than half. For example, you could take $a = 1/4$ and $b = 3/4$.

Now pick three numbers, a, b, and c, so that all lie in separate thirds (that is to say, one is less than one-third, another is between one-third and two-thirds, and the third is bigger than two-thirds). This too is easy. You can use the same values of a and b as before and pick c equal to 1/2.

Then the problem asks us to pick four numbers, a, b, c, and d. We have to do this so that a and b are in separate halves, a, b, and c are in separate thirds, and a, b, c, and d are in separate quarters. This requires a bit more effort. We cannot use any of our previous choices of a, b, c, since these will not be in separate quarters. Instead, we have to start again with four new numbers. Say: $a = 1/8$, $b = 7/8$, $c = 3/8$, and $d = 5/8$.

Next we have to find five numbers, the first two in separate halves, the first three in separate thirds, the first four in separate quarters, and all five in separate fifths. And so on.

After you have done a couple of these, you will soon get the hang of it. Though you sometimes have to start afresh with a whole new batch of numbers, there does not seem to be any reason why you could not, in principle, go on for ever.

At this point in his explanation, George paused to look up, a challenging smile on his face. "Doesn't there?" he asked.

As the mathematician in the party, I saw— or thought I saw—the subtle point that the question was about. I knew, as will every other professional mathematician, that there are two quite different things that could be meant by that phrase "go on for ever."

It could mean that for every whole number N, however large, you can find N fractions $x_1, \ldots, x_N$ so that the first two are in separate halves, the first three in separate thirds, and so on, with all N numbers being in separate Nths.

Or it could mean that you can pick a single, infinite sequence of fractions, $x_1, x_2, x_3, \ldots$ and so on, ad infinitum, so that the first two are in separate halves, the first three in separate thirds, the first four in separate fourths, and so on.

The difference between these two possibilities is essentially this. In order to find an infinite sequence as in the second case, you have to be able to pick x_1, x_2, x_3, and so on, in turn, once and for all, without coming back to change your mind. In the first case, for each new value of N you can start all over again if you like.

Having seen how to handle the first few cases, I was sure I knew the answer to George's as yet unspoken question. Obviously, I said, you can always find a solution in the first case (given the N, pick the N fractions), but (and here I was less sure, but 25 years worth of mathematical intuition seemed to point in this direction) you cannot pick a single infinite sequence of numbers that works as in the second case.

(Incidentally, this has nothing to do with actually writing down an infinite sequence of numbers. All that would be required in the second case is to describe the procedure by which you could produce all your numbers one after another.)

George waited to see if I wanted to change my mind. No, I was sure I was correct. It was obvious you could go on forever, getting

longer and longer sequences of numbers, and I was reasonably confident that given an hour or so on the plane to San Francisco the next morning, I would be able to prove that the infinite sequence version was not possible.

The others around the table, none of them mathematicians, looked suitably impressed. And I basked in my glory . . . until George told us the real answer.

For according to George, you can keep going as far as a sequence of 17 fractions, but it is impossible to find 18 fractions that divide up in the appropriate manner. It was not until I received a copy of the solution that I believed this result.

The problem appears to have been raised first by the mathematician Steinhaus, who gave it, along with the solution, in his 1964 book *One Hundred Problems in Elementary Mathematics* (Basic Books). Another solution was obtained by Berlekamp and Graham in 1970.

The simplest solution, and the one George sent me, was found by M. Warmus of the Computation Center of the Polish Academy of Sciences in Warsaw, and published in the *Journal of Number Theory,* Volume 8 (1976), pp. 260–263.

Unfortunately, being three pages long, the solution does not lend itself for reproduction here, but it requires no advanced mathematics. An evening spent with a pencil and paper should enable you to come up with 17 numbers that work. To show that you cannot get 18 numbers, the trick is to look first at that number in such a sequence that would have to lie between 4/9 and 5/9.

Good hunting. But be warned that even if you sent me an elegant solution, I think I will still not really believe the answer. Mathematics just isn't like that. Is it?

125

The vertical confusions

April 27, 1989

The story is a familiar one to all regular readers of the *Micromaths* column. In a note scribbled in the margin of a textbook and discovered after his death, the great seventeenth-century French mathematician Pierre de Fermat claimed that he had found a proof of the fact that for all values of n greater than 2, the three-variable equation:

$$x^n + y^n = z^n$$

has no positive whole number solutions (except for all three variables being equal to zero). Unfortunately, Fermat did not include any details of his claimed proof, leaving the world with merely the tantalizing comment that it was (a) "truly marvellous," and (b) "too large to fit into the margin."

Thus was Fermat's Last Theorem born into the world. Since then it has resisted thousands of attempts at solution, ranging from the valiant lone attempts by amateur mathematicians to the full frontal assaults of the sophisticated professional.

There is little wonder that the problem has attracted so much attention: it was posed by one of the most famous mathematicians of all time; the problem is ludicrously simple to state, requiring no deep knowledge of mathematics to understand; no one has been able to solve it in over 300 years; work on the problem has led to some significant developments in mathematics; and on top of all that, there is a large cash prize awaiting the first person to find a correct solution.

Computer-aided work has demonstrated the validity of Fermat's claim for all values of the exponent n up to 125,000 and as recently as 1983, a young West German mathematician managed to prove that for any value of n, there can be at most a finite number of (essentially different) solutions for x, y, z. ("Essentially different" means ignoring simple multiples of solutions.) Not very much else is known, and the problem continues to tantalize the professional and amateur alike.

All the evidence, both that gleaned from an examination of the rest of Fermat's mathematical work, and the accumulated weight of 300 years worth of failures, points to the conclusion that the great Frenchman was in fact mistaken in his claim—that his "truly marvellous proof" contained a serious flaw.

Indeed, there are grounds for concluding that Fermat himself subsequently realized his error. Certainly that one tantalizing note in the margin, not intended for publication, was the only reference he made to the problem in any of his known writings.

But there is another possibility that resurfaces every few years, to the effect that maybe Fermat's marginal claim was made not as a result of a mathematical error, but rather as a result of a rare form of dyslexia, whereby letters or numbers become confused vertically.

Perhaps, so this story goes, the equation Fermat meant to write down was not the one given above, but rather the equation you get by vertically interchanging the variables and the exponent, namely:

$$n^x + n^y = n^z.$$

For this equation, it is indeed possible to find a truly marvellous proof that there is no solution (that is, no positive whole numbers x, y, z) for values of n greater than 2. (For $n = 2$, there are infinitely many solutions, just as with the more usual "Fermat equation.")

So was this the explanation all along? Probably not, but this does not detract from the fact that the dyslexic version to the Last Theorem is indeed a "truly marvellous" little problem to solve.

This is the chance that all *Micromaths* readers have been waiting for. Proving Fermat's Last Theorem seems out of the question, but at least you can obtain the satisfaction of proving the dyslexic version.

The proof does require some careful thought, but you should be encouraged by the fact that it is short enough to fit in the margin of this newspaper.

126

Introducing the figure that always adds up
May 18, 1989

Pick any number consisting of three different digits. You are allowed to use zero, even as the first digit if you like. Now reverse the order of this number and subtract the smaller of your two numbers from the larger. Reverse the digits of that number and add it to your previous answer. The result will be 1089. Always.

Suppose you start with the number 125. Reversing this gives 521. Subtract 125 from 521 to get 396. Reverse this to get 693. Add 396 to 693 to obtain 1089.

Again, suppose you start with 372. Reversing gives 273. Subtracting the smaller from the larger gives 099. (Note the leading zero to ensure that you still have a "three-digit" number.) Reversing this gives 990, and adding 099 to 990 gives 1089.

Another intriguing property of the number 1089 is provided by Table 1, where reversing the digits in each of the products of 1089 by 1 through to 9 results in the same sequence of products in reverse order.

In fact, a closer look at the table will reveal another curiosity. If you look down each of the four vertical columns of digits in the last two columns of four-digit numbers, you will see one of the two sequences 0 through 8 or 1 through 9 in either ascending or descending order.

Mathematics abounds with patterns and symmetries of this form, and all it takes to discover them is some paper and a pencil

Table 1

$1089 \times 1 = 1089$	reverse	9801
$1089 \times 2 = 2178$	reverse	8712
$1089 \times 3 = 3267$	reverse	7623
$1089 \times 4 = 4356$	reverse	6534
$1089 \times 5 = 5445$	reverse	5445
$1089 \times 6 = 6534$	reverse	4356
$1089 \times 7 = 7623$	reverse	3267
$1089 \times 8 = 8712$	reverse	2178
$1089 \times 9 = 9801$	reverse	1089

and a little patience. If you do, you will be in good company. In his celebrated *Life of Samuel Johnson,* written in 1791, Boswell wrote, "When Dr. Johnson felt, or fancied he felt, his fancy disordered, his constant recurrence was to the study of arithmetic."

To start you off, see what happens when you construct a table of all the products 7×7, 67×67, 667×667, 6667×6667, and so on. Then take a look at a similar table consisting of the squares of the numbers 4, 34, 334, 3334, etc. And how about the squares of the numbers 9, 99, 999, 9999, and so on?

Or, for something of a different nature, see what you get when you multiply the number 15873 by each of the numbers 7, 14, 21, and 28 (that is, multiples of 7).

And then, when you have tried all those, take off on your own on a journey of discov-

ery. Perhaps you will find a pattern as beautiful as the one in Table 2, my all-time favorite
number pattern.

Table 2

1	×	1	=	1
11	×	11	=	121
111	×	111	=	12321
1111	×	1111	=	1234321
11111	×	11111	=	123454321
111111	×	111111	=	12345654321
1111111	×	1111111	=	1234567654321
11111111	×	11111111	=	123456787654321
111111111	×	111111111	=	12345678987654321

127

Get Knotted

June 1, 1989

Imagine that the figure shown in part (i) of the picture is made of perfectly stretchable elastic. You have the task of deforming the figure in (i) so that it looks like the one shown in (iii). Can it be done?

Most people would say it's impossible, unless you cut one of the interlocking rings, as indicated in part (ii), then rejoin the two cut ends.

But in fact it is possible to deform the figure from (i) to (iii) using manipulation alone,

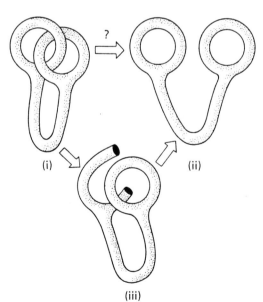

(i) (ii) (iii)

FIGURE 35. A knotty problem. Can you see how to transform (i) into (ii) without cutting as shown in (iii). Imagine the figure is made out of perfectly stretchable material.

without any cutting. How? Well, that is for you to find out. The solution will appear in the next *Micromaths*. [Chapter 128.]

This tantalizing little puzzle is an example of the kind of problem that arises in the study of topology, the mathematics of perfectly elastic objects. Though they look quite different, the two figures (i) and (iii) are in fact different ways of looking at essentially the same object. They are known as "topologically equivalent."

As the puzzle itself shows, it can be extremely difficult to visualize manipulations of knotted figures and in particular it is nigh on impossible to be sure that two seemingly different figures are in fact topologically equivalent.

An even more difficult task is to show that two seemingly different figures are in fact topologically *different,* that is to say, cannot be manipulated the one into the other. In the case of topological equivalence, once you have found the right manipulations, at least you are then sure the two figures are topologically equivalent. But in demonstrating topological non-equivalence, things are much harder. No matter how many manipulations you try and show to fail, how could you ever be sure there is not some clever manipulation you have missed?

So how can you show that two figures are topologically non-equivalent, if the obvious method of trying manipulations does not work? The answer is to use mathematics.

The mathematical study of knots was begun by Listing, a student of the legendary Gauss, back in the 19th century. The only special stipulation required for a mathematical knot is that the two otherwise free ends of the knotted cord be joined together, so the entire length of the cord forms a knotted, closed loop.

The reason for this last requirement is obvious when you stop to think about it. The main problem in knot theory is how to decide when two seemingly different knots are in fact the same; that is to say, when they are topologically equivalent. Thus, we imagine the two knots made out of perfectly stretchable string, and ask if it is possible to deform one knot into the other.

Clearly, if the ends of each knot are free, we can simply untie one knot and then retie it in the form of the second. So the mathematician gets around this problem by always demanding that knots have no free ends.

Over the years, knot theorists have developed an array of techniques for demonstrating that similar knots are in fact different. The easiest to use are the so-called "knot polynomials," the first of which were discovered by Alexander in 1928.

Alexander showed how to associate each knot with a certain polynomial in one unknown, x, in such a way that if two knots have different Alexander polynomials, then the knots are topologically inequivalent.

For example, for the sealed-end version of the common overhead knot, the Alexander polynomial is

$$x^2 - x + 1$$

and for the sealed-end version of the figure-eight knot it is

$$x^2 - 3x + 1$$

Since these two polynomials are different, it follows that the two knots are topologically nonequivalent.

Though simple to use, the main drawback with the Alexander polynomials is that they do not always work. It is possible to have different knots that have the same Alexander polynomial. For instance, the familiar reef and granny knots both run out to have the same Alexander polynomial,

$$x^4 - 2x^2 + 1$$

so the Alexander polynomials fail to do what the average boy scout can accomplish with ease.

It was not until 1984 that this state of affairs was rectified, in that a more complicated class of polynomials was discovered that were able to tell the reef knot from the granny, and indeed most known pairs of knots for which the Alexander polynomials failed. They are known as Jones polynomials, in honor of the man who discovered them, mathematician Vaughan Jones of the University of California at Berkeley.

128

Tarski and hunch squares the circle

June 29, 1989

So now we know it is possible to square the circle. I do not mean that someone has finally found a solution to the ancient Greek challenge to construct a square with area equal to that of a given circle using ruler and compasses only. The continued, valiant efforts of would-be circle-squarers around the world notwithstanding, the 1882 proof that it is an absolute impossibility to square the circle that way remains intact.

But a Hungarian mathematician has found a solution to another circle-squaring problem —one posed by the famous mathematician and logician Alfred Tarski in 1925. Tarski asked if it was possible to partition a circle (including its interior) into finitely many sets that can be rearranged to form a square of the same area.

"Yes," announced Miklos Laczkovich earlier this year. And after examining his 40-page proof, other mathematicians agree. However, the layman should be cautioned that the proof is highly technical, and does not provide a method whereby one could take a pair of scissors and turn a circle into a square.

In fact, Laczkovich's circle must be cut into about 10^{50} pieces, in bizarre shapes that in mathematician's parlance are known as "non-measurable sets."

Although the number of pieces is large (10^{50} is the order of the number of molecules of water in a million cubic miles of ocean),

the fact that it is finite means mathematicians must rethink their understanding of what it is to be curved. For the idea that any finite number of cuts, no matter how large and how effected, can turn a circle into a straight-edged square, defies the imagination of even the most sophisticated mathematician.

Another such oddity, a precursor of the new result, is its three-dimensional analogue, known as the Banach–Tarski Paradox. This says it is possible to take a solid ball and cut it into a finite number of pieces that can be reassembled to form a cube of the same volume.

Though superficially as hard to believe as the squaring of the circle, the Banach–Tarski result is not particularly hard to prove, at least not by the standards of modern mathematics.

Because of the way "volume" (more precisely, "measure") behaves in three dimensions, it turns out that a sphere can be decomposed to form a cube of any volume whatsoever, or, for that matter, two new spheres each equal in size to the original sphere. Though this result is known as a paradox, it is very well understood. But there were good reasons to suppose that what happens in three dimensions cannot occur in two. For one thing, "area" (that is, "measure") in two dimensions behaves quite differently from "volume" ("measure") in three.

In particular, it is impossible to have a two-dimensional analogue of the Banach–Tarski

Paradox, whereby a circle is decomposed into any other figure having a different area.

Prior to the new discovery, some two-dimensional results were known, but the evidence they provided suggested that the circle could not be squared. Certainly, any polygon could be decomposed into a square (of the same area). This easy result was proved by Tarski himself back in 1924. But polygons do not have the curved edge that appeared to block a similar decomposition of the circle into a square. In fact in 1963, mathematicians Lester Dubins, Morris Hirsch, and Jack Karush proved conclusively that the circle could not be squared by any cutting procedure that dissected it into pieces that have any chance of being recognizable to the naked eye.

Then, in 1985, Richard Gardner went even further, demonstrating that circle-squaring is impossible even when certain more bizarre forms of cut are allowed. Following Gard-

ner's result, most mathematicians would have guessed that the problem is as impossible as its older, and more famous Greek namesake—except, it seems, Miklos Laczkovich of the Eötvös Lorand University in Budapest, who was inspired to work on the problem after attending a lecture given by Gardner in Italy.

For all its counter-intuitiveness, and despite the evidence to the contrary, the circle can be squared in the Tarski sense. It leaves the layman as baffled as ever at the strange and mysterious world of higher mathematics, and presents the mathematician with some serious rethinking of what it means for a line to be curved.

Finally, here is the solution to the knotty problem posed in the last *Micromaths* [Chapter 127]. The diagram shows some key intermediate steps involved in deforming figure (i) to the figure (v). So now you know.

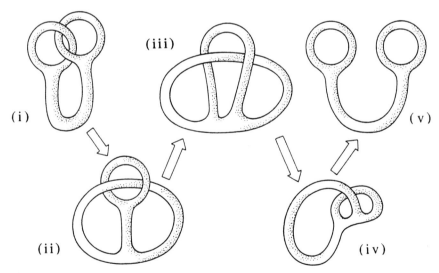

FIGURE 36.   The knot problem unravelled. There are no cuts, but the solution stretches the imagination as much as it does the knotted figure.

129

Today's moment of history

July 6, 1989

Shortly before half-past one this afternoon, you might like to pause and look at your watch, preferably a modern-day, digital variety. For at precisely 23 minutes and 45 seconds after the hour of 1 o'clock, an unusual event takes place. Today being July 6, 1989, at that very instant you are at the moment of time whose complete numerical description is 1.23.45.6.7.89—all the decimal digits in their correct order. Clearly, it's a unique moment in history for those interested in numerology.

In fact, had you been both awake and aware of this fact at the time, at 1.23.45 this morning the numerical description of that instant in time, using the 24-hour clock, was 01.23.45.6.7.89, which includes the number 0 in its correct position. Even better.

If you miss this historic instant, then at least next year you get a chance of a moment almost as good. At 56 seconds past 12.34 on August 7, 1990, the time will have the description 12.34.56.7.8.90. With the 0 occurring at the end of this numeric sequence, you cannot make use of the 24-hour clock to get two bites at the cherry, as was the case this year.

After that, there is a long wait for a similar event, which will not occur until the next century draws to a close, bringing with it 2089 and 2090. By the time these two years come around, mention of the prefix "20" will doubtless have long ceased, with years referred to by their final two digits as is now commonly

the case, so the same digit-sequence phenomenon will occur.

And with the turn of the century comes another date of mathematical interest, in this case one of considerably more significance than today's numerical curiosity. Though people will inevitably find it a novelty to start numbering the years 00, 01, and so on, one particular group of people face an enormous and potentially-catastrophic problem. For data processing managers and their teams of programmers, the dawn of the next century poses a significant challenge.

Most of the computer programs that keep track of bank accounts, credit card payments, utility bills, school records and so on refer to the current year by the last two digits, just as we do ourselves. As one year gives way to the next, the two-digit number that keeps track of the years increases by one. That's simple and straightforward.

But when 1999 turns into 2000 and the computer's reference number clicks from 99 to 00, what then? In many cases the answer is that no one really knows. Large database systems are so complex, and in most instances have been subjected to so many amendments over the years, that the effect of a double-zero year number can be impossible to predict. A computer that thinks years are measured using two-digit numbers may not recognize that the world continues after year 99, with the count restarting at 00.

In practical terms the only hope is that before this century is out, most of the computer systems currently in use will have been phased out and replaced by newer systems. Present day programmers are (or at least certainly should be) aware of the approaching century-end, and making sure that their programs are able to recognize that the year 00 is the next one after 99. But just in case, the prudent citizen would be well advised to retain all their bank balance sheets and utility bills as the century draws to a close.

With the news that today is mathematically unique, *Micromaths* is about to take a short summer vacation. With my two-year stay at Stanford University coming to an end, I am about to hit the great American freeway system, before settling back in front of my Macintosh in a few weeks time.

Finally, David and Gregory Chudnovsky of Columbia University, New York, have just computed the mathematical constant π to a record-breaking 480 million decimal digits. This easily eclipses the previous record of 201 million places, held by Yasumasa Kanada of Tokyo University since 1986 (and reported here at the time).

The Chudnovskys performed this essentially useless calculation twice, once on an IBM 3090 mainframe, and then again on a Cray-2 supercomputer, thereby ensuring that not a single one of the 401 million digits is incorrect. This reassuring fact will enable us all to sleep soundly in our beds tonight.

Postscript.   See Chapter 131 for an update on the calculation of π.

130

Playing it by numbers
October 12, 1989

Micromaths returns from its summer break with the news that the four-year-old record for the largest prime number has just been shattered.

In the early hours of August 6, an Amdahl 1200E mainframe carried out a 33-minute calculation that confirmed the prime number status of the 65,087-digit number

$$391,581 \times 2^{216193} - 1.$$

This giant of a number is some 37 digits longer than the previous record, set in 1985 by David Slowinski of Cray Research, using a Cray XMP supercomputer [Chapter 63].

Sharing the honors for the new discovery is a team of Amdahl researchers: John Brown, Landon Curt Noll, B.K. Parady, Gene Smith, Joel Smith and Sergio Zarantonello. Their result brings to an end a string of prime number records obtained on Cray computers, though Cray remains in the picture with the independent verification of the Amdahl number being carried out by Cray researcher Jeff Young.

For the uninitiated, a prime number is a whole number that can only be evenly divided by itself and 1. Examples are 2, 3, 5, 7, 11, 13, 17, 19—the prime numbers less than 20. With the exception of 2, all prime numbers are odd.

This list of primes might suggest that most odd numbers are prime. This is not the case. The further away from small numbers you get, the rarer the prime numbers become.

But though the prime numbers "thin out" as you look at larger and larger numbers, they never stop. It has been known since Euclid's time that there are infinitely many prime numbers. This provides the number-curious among us with an irresistible challenge: find a prime number bigger than all others known to man.

A few moments with a pencil and paper should be enough to convince you that the task cries out for a computer attack. But you should realize that this game has long since moved beyond the sphere of the home-computing enthusiast. A combination of clever mathematics and the most powerful computers are required to deal with numbers of the magnitude of the record primes.

In fact, the last time an "amateur" discovered a record prime number was in 1979, when 19-year-old Curt Noll used a California State University computer to discover the then-record 6,987-digit prime number $2^{23209} - 1$.

And this same Noll was part of the Amdahl team that made the latest discovery. Noll's earlier success, like that of all other recent prime records until now, involved a search among the so-called Mersenne numbers: Numbers of the general form $2^N - 1$, where N is a whole number.

It is not that such numbers are likely to be prime numbers. In fact, prime Mersenne numbers are extremely rare, with only 30 of them known to date against billions of "ordinary" primes. However, a very powerful method, known as the Lucas–Lehmer Test, makes the prime-number status of Mersenne numbers a relatively easy matter to test.

The Amdahl prime is not of this form, but of a related form which allows the use of a version of the Lucas–Lehmer Test as modified by Hans Riesel. This, and a veritable arsenal of ingenious computational tricks, enabled the Amdahl team to make its discovery.

What use is it? To the mathematician, it is important to know that the primes are infinite in number. And the identification of 100-digit primes plays a crucial role in modern cryptography. But knowing that a particular number with thousands of digits is prime is of no importance whatsoever.

What is relevant (and this explains why otherwise profit-conscious organizations become involved in such record attempts) is the development of the computers and the mathematical algorithms capable of handling such large numbers.

Discovering a record prime not only tests such machinery and methods to their limits, it usually attracts sufficient publicity to ensure that everyone knows you have developed such a capability.

The appearance of this article shows this is indeed the case.

131

Pi-eyed over eternal sum

November 16, 1989

The one-billionth digit of π is 9, and that's official. It comes to you courtesy of Gregory and David Chudnovsky, two professors at Columbia University, New York, who recently completed the calculation of the mathematical constant π to a record-breaking 1,011,196,691 decimal places.

The previous record holder in the game of π-mania was a Japanese team led by Yoshiaki Kanada, who in 1988 calculated it to 201,000,000 places. [Chapter 77.]

Most of us remember that π is 22/7, but this is just an approximation. The number π is the result you obtain when you divide the circumference of any circle by its diameter. Since the last century, it has been known that this number cannot be expressed exactly using standard decimals. If you were to try to write out the decimal expression for π, you would never complete the task.

A better approximation to π than 22/7 is the figure 3.14159 26535, which is π to 10 decimal places. So what would a billion decimals places of π look like? Well, imagine *The Guardian* devoting one day's issue to the publication of the first billion digits of π. Using standard type, we could print 30 digits on a line across eight columns per page, with 185 lines per column. A single page would therefore contain some 44,400 digits. To print the entire billion digit expression would require about 22,500 pages. Your *Guardian* would be about 30 inches thick! And there would be no news.

Calculation of π has long been a popular pastime among the numerically minded. To perform such a computation, mathematicians do not start out by measuring circumferences of circles. Rather they make use of one of a number of mathematical formulas, the oldest of which is due to Liebniz in the 17th century:

$$\frac{\pi}{4} = 1 - \frac{1}{3} + \frac{1}{5} - \frac{1}{7} + \frac{1}{9} - \cdots .$$

The dots at the end of this sum mean that the successive additions and subtractions continue for ever. Of course, it is not possible to complete such a sum, but since the terms being added and subtracted grow smaller, calculating more and more terms of the infinite sum gives successively better approximate answers.

A mathematical analysis of the infinite sum can be used to indicate how many terms need to be added to obtain π to however many decimal places of accuracy are required. In fact, the sum quoted is not a very good one to use. The individual terms grow small so slowly that you need to include a lot of them to achieve any reasonable accuracy when calculating large numbers of decimal places.

The latest computation makes use of an infinite sum for π discovered by the Indian mathematician Srinivasa Ramanujan, which produces large numbers of decimal places relatively quickly.

Even so, to get a billion places, you need far more than Ramanujan's clever formula (which is too complicated to print here). You need the latest in computer technology and some very ingenious computer mathematics.

In particular, the Chudnovskys made extensive use of a powerful multiplication technique known as the Fast Fourier Transform, which they tailored specially to the problem at hand. The program was written in regular old Fortran, and the computation was performed on two machines, a Cray 2 supercomputer at the Minnesota Supercomputer Centre and an IBM 3090 mainframe at the IBM Research Center in Yorktown Heights, New York.

What use is the new result? None whatsoever. Such computations serve simply as benchmarks for computer technology, both hardware and software, and in particular the very efficient algorithms mathematicians have devised in order to handle such immense calculations.

And a billion decimal places of π do not come by brute computing power alone. Indeed, the Chudnovskys claim that their methods are sufficiently powerful to enable still more decimal places to be computed using fairly standard computing equipment.

So things have come a long way since the days of poor old William Shanks. In the 19th century, this gloriously eccentric Englishman devoted 20 years of his life to the calculation of π to 707 decimal places, publishing his result in the *Proceedings of the Royal Society* in 1873–74. Unfortunately, Shanks make a mistake in the 527th and subsequent decimal places, so a lot of his labors were in vain. But this was not discovered until 1945 with the advent of desk calculators, by which time Shanks himself was long past caring.

Still, Shanks' correct 526 decimal places were a huge improvement on the value of 3.125 obtained by the Babylonians some 4,000 years earlier, or the value of 3 implied by certain parts of the Bible (I Kings 7:23 and II Chronicles 4:2).

132

Their infinite wisdom

November 30, 1989

"Infinity is a fathomless gulf, into which all things vanish." So wrote the Roman Emperor Marcus Aurelius in the second century. Even more evocative is John Milton's "Infinity is a dark illimitable ocean, without bound." Then there is the oft-repeated schoolroom definition: "Infinity is where things happen that don't," such as the meeting of parallel lines.

But, more pertinent to this column, what does the mathematician mean by "infinity"?

To the mathematician, infinity is the keystone that supports everything else—or at least the greater part of everything else. Indeed, large parts of modern mathematics consist of nothing else than a systematic way to handle the infinite.

The high-school calculus course that forms the gateway to all higher mathematics is simply a collection of methods for dealing with the infinitely small and the infinitely large.

The mathematician's infinity arose with the ancient Greeks. Though unable to develop the mathematical machinery to handle infinity, they recognized the problems that it brought.

Zeno illustrated them with a series of famous paradoxes, one of them being the paradox which purports to demonstrate that motion is impossible.

Zeno's argument goes like this. Imagine a runner who sets out to run a mile race. Before he can cover the entire mile, he must first cover half that distance. Then he has to cover half the remaining half-mile. Then half the remaining quarter-mile. Then half the distance still remaining, and so on. Since this requirement of having first to cover half the remaining distance continues indefinitely, the runner can never reach the finish.

Of course, Zeno knew full well that the runner would reach the finish. The paradox lies in trying to explain what is wrong with the logic that says it is impossible to satisfy an infinite sequence of "prior conditions."

Modern mathematics easily disposes of Zeno's problem. The requirements of first covering half the remaining distance translates to the runner having to cover the distance represented by the infinite sum

$$\frac{1}{2} + \frac{1}{4} + \frac{1}{8} + \frac{1}{16} + \cdots .$$

If you let S stand for the answer to this sum, then multiplying the entire sum through by $\frac{1}{2}$ gives the same sum with the first term missing:

$$\frac{1}{2}S = \frac{1}{4} + \frac{1}{8} + \frac{1}{16} + \frac{1}{32} + \cdots .$$

Subtracting the second sum from the first, all of the terms cancel out except for that initial $\frac{1}{2}$ in the first sum, which produces the result

$$S - \frac{1}{2}S = \frac{1}{2}$$

which means that $S = 1$.

But is it permissible to multiply through an infinite sum and subtract in this way? After all, an infinite sum can never be worked out the way a finite sum can (in principle, given enough time and computing power).

What works for finite sums does not necessarily work for infinite ones. Look what happens with the infinite sum

$$1 - 1 + 1 - 1 + \cdots .$$

If you let S represent the answer, then multiplying the sum through by -1 gives the same sum minus the first term, so

$$-S = -1 + 1 - 1 + 1 - 1 + 1 - \cdots .$$

Now if you subtract the second sum from the first, all the terms cancel out except for the initial 1 in the first, and you get the equation

$$S - (-S) = 1$$

so

$$S = \frac{1}{2}.$$

But wait a moment. The original sum consists of an infinite sequence of pairs of subtractions $1 - 1$. That is:

$$(1 - 1) + (1 - 1) + (1 - 1) + (1 - 1) + \cdots .$$

Surely this can be written as

$$0 + 0 + 0 + 0 + 0 + \cdots$$

which is 0, is it not?

Or, if you remember that two minus signs make a plus, you can rewrite the sum as

$$1 - (1 - 1) - (1 - 1) - (1 - 1) - \cdots$$

and this time you get the answer 1. So what is the right answer, $\frac{1}{2}$, 0, or 1? The right answer, in fact, is that this particular infinite sum does not have an answer.

This immediately raises the question as to how can you tell when an infinite sum has an answer, and when it is permissible to handle infinite sums in the same way as finite ones.

You might think that the distinction lies in the fact that in the first sum you keep adding increasingly smaller numbers, but in the second the numbers don't get smaller, they just keep changing sign.

This is part of the story, but by no means all of it. For instance, an extremely important infinite sum is the so-called harmonic sum:

$$1 + \frac{1}{2} + \frac{1}{3} + \frac{1}{4} + \frac{1}{5} + \frac{1}{6} + \cdots .$$

Though the numbers in this sum get progressively smaller, the sum does not have an answer. Or at least, the answer it does have is "infinity." But that seems to be conferring on "infinity" the status of some kind of number—like 1, 2 and 3, perhaps, only bigger. And there begins a whole new chapter of mathematics.

133

Taming infinity

December 14, 1989

Gabriel's Horn is a musical instrument that exists only in the minds of mathematicians. With its mouthpiece in heaven, its sound emerges at the trumpet-end an infinite distance away on Earth. But for all its practical impossibility, this hypothetical instrument serves to illustrate some of the bizarre things that arise when you start to investigate the infinite.

Last time [Chapter 132], *Micromaths* stepped boldly into the world of the infinite with a discussion of how infinity underlies practically all of present-day mathematics. The ubiquitous calculus was cited as an example of mathematicians' attempts to tame the unruly infinite. And it is the calculus that provides the most surprising property of Gabriel's Horn.

To construct the horn, the mathematician takes the graph $y = 1/x$ for values of x from 1

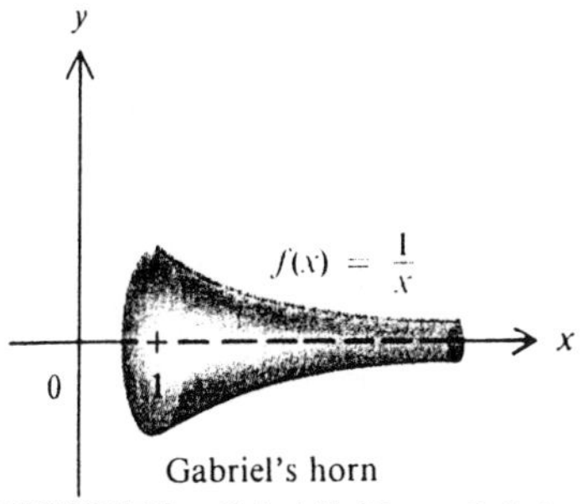

FIGURE 37. Gabriel's Horn. Infinite area, finite volume.

up through all larger numbers (i.e., "up to infinity") and imagines this infinitely-long graph rotated about the x-axis to form an infinitely long trumpet, whose opening is at the point $x = 1$ and whose mouthpiece is located far away at infinity.

Though impossible to construct in real life, techniques known since Newton's time are perfectly adequate to deal with this object in a mathematical fashion—techniques familiar to anyone who has covered high-school calculus. The method of integration allows the mathematician to calculate the volume of Gabriel's Horn and its surface area.

The volume turns out to be nothing other than the old standby, the mathematical constant π. The surface area of the horn works out to be infinity. That might seem obvious. After all, why should it not be the case that an infinitely long trumpet has an infinite surface area, and would therefore require an infinite amount of sheet metal to manufacture?

An amount π of paint would exactly fill the trumpet so that the entire interior surface was in contact with paint. And yet, having an infinite area, it ought surely to require an infinite amount of paint to coat the interior surface? How then can a finite amount (π) of paint fill the trumpet?

Of course, the paint would have to be mathematical paint that could be applied in an infinitesimally thin layer. Nevertheless, this peculiar property of Gabriel's Horn does serve to illustrate what a bizarre world awaits those who start to tamper with the infinite.

Hilbert's Hotel provides another example of how strange the infinite can be. Named after mathematician David Hilbert, this hotel-to-end-all-hotels has infinitely many rooms which are numbered $1, 2, 3, 4 \ldots$ in the usual way. One night you arrive at the hotel only to find that all the rooms are taken. (Hilbert's Hotel is located in the world of mathematics, where there are infinitely many people.)

Under normal (i.e., finite) circumstances this would mean that you would have to find another hotel. Not so with Hilbert's Hotel. You simply ask the desk-clerk to move everybody to the next room. The occupant of room 1 moves to room 2, the person in 2 moves into 3, and so on, all the way through the hotel. This leaves room 1 empty, and you have a bed for the night. And though everyone has had to move into the next room, no one has had to leave. Everyone is happy, especially the hotelier whose profits, already infinite, have increased even more with your arrival.

In fact, the hotelier can double his takings should it happen that infinitely more new guests turn up. It is possible to move all the present occupants into new rooms so as to accommodate infinitely many new arrivals. Can you figure out how to do it?

With the infinite behaving in such a bizarre fashion, it is scarcely surprising that infinite numbers have an arithmetic quite unlike the arithmetic of finite numbers.

Infinite numbers? Plural? Surely, if there is an infinity at all, there is just one of them.

Again, when you start to look in more detail, you soon discover, as did George Cantor at the turn of the century, that there are infinitely many different infinities. The infi-

nite numbers come one after the other, getting steadily bigger, just like the regular numbers.

Though you might feel uneasy about the very idea of infinite numbers, arithmetic for such creatures turns out to be much less of a chore than for the finite numbers. If you add together two infinite numbers, the result is just the larger of the two numbers. So, too, if you multiply two infinite numbers: the result is equal to the larger of the two. And as for subtraction and division, well, these don't arise at all. You cannot subtract or divide infinite numbers. It's a schoolboy's dream.

Or is it rather a nightmare? Taming the infinite has taken mankind many centuries. One of the triumphs of 19th century mathematics was the development of a proper mathematical theory to underpin Newton's calculus, and that was achieved by finding skillful ways to avoid infinity as much as possible.

It was only some 30 years ago that the late Abraham Robinson found a way to do calculus that really came to grips with the infinite. Understanding either theory, the 19th century one or Robinson's, provides today's university student of mathematics with an enormous challenge.

Many fail fully to understand the way infinity behaves. For example, I recently posed the following question to a class of very bright mathematics undergraduates. Think of the number $0.9999\ldots$ (with a 0 before the decimal point and an infinite sequence of 9s after the point). Is this number truly equal to 1 or is it just "very close" to 1? Almost two-thirds of the class said it was almost equal to 1, but not quite. They were wrong. $0.9999\ldots$ is completely and utterly equal to 1. But I'll leave you to figure out a proof of that fact.

- *Time*'s 1990 Men of the Year:
  The 2 George Bushes
- *Times*'s 1991 Man of the Year:
  Ted Turner
- *Times*'s 1992 Man of the Year:
  Bill Clinton
- *Time*'s 1993 Men of the Year:
  Yitzhak Rabin, Nelson Mandela,
  F. W. de Klerk and Yasser Arafat
- US invades Panama
- Soviet Union gets first
  McDonald's
- Bush HATES broccoli
- Jack Kevorkian, MD
- Michael "junk bond
  king" Milkin gets
  10 years
- Thatcher resigns
- Shelby Foote & Ken Burns
  bring *The Civil War* to PBS
- *In Living Color*
- Desert Shield
- Desert Storm
- Anita Hill and Clarence Thomas
- "Magic" Johnson has AIDS
- Fires in Kuwait
- Bosnia
- Norman Schwarzkopf
- Brady Bill
- A chance for peace in the Middle East (?)
- Elections in
  South Africa

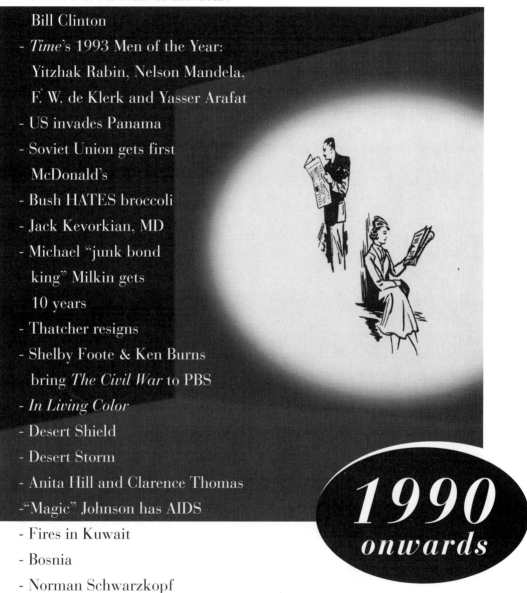

1990
onwards

134

Call to order

January 25, 1990

On Monday, January 15, I was at a mathematics meeting at the University of California at Berkeley. The following day I was due to travel on to a second meeting in Louisville, Kentucky. So I did what every seasoned traveler does: called ahead to check my flight reservations and hotel booking.

At least I tried to, but I was unable to get through either to the airline or the hotel. At the same time, across the US, countless other travelers were having the same difficulty. So too were journalists trying to phone through their stories or connect their computers to one of the news databases.

Though I did not know it at the time, something had gone seriously wrong with the usually reliable AT&T phone system. The following day's *New York Times* front page told me: an error in the computer program that controls the entire telephone network had crippled the system.

By good fortune, it had happened on a public holiday, Martin Luther King Day, so the network did not have to handle its usual 80 million calls. Even so, for a nine-hour period starting just after 2:30 P.M., over half the calls attempted failed to get through.

A technician at AT&T's Bedminster, New Jersey, network operations center described the scene there as they tried frantically to track down the problem as "chaotic and almost out of control." The president of AT&T described it as the most serious networking problem in the company's history.

Technical experts from the nearby Bell Laboratories (the research wing of AT&T) had to be called in to figure out an emergency software repair, but by evening there were still long delays in calls being connected.

The last time there had been such severe disruption to the vast US telephone network was in 1988. On that occasion the problem had been rather more prosaic. A construction crew had inadvertently cut through a fibre-optic cable.

The latest problem stemmed from an attempt to improve the performance of the system. Unfortunately, as *Micromaths* has pointed out on numerous occasions, it is simply not possible to write complex software without making any errors. The surprising thing is that so much software does in fact work pretty well. But this time one particular bug did surface, with devastating effects.

The computer programs that are used to control today's nationwide telephone networks are probably the most complex software systems ever designed, involving many thousands of lines of code. When a caller in, say, Berkeley, California, tries to call an airline in Chicago (as I did on the electronically Black Monday), the main control program has to figure out which of the millions of possible routing combinations is the most economical, check whether all the intermediate connections on that route are available, and then ensure that the phone at the receiving end is able to receive the call.

In the case of a call made using a telephone credit card (as mine was), the system also has to refer to an on-line database to check that the password number the caller punches in is valid. The system then has to start to time the call, figure out the charge-rate and arrange for the cost of the call to be billed to the correct account.

Some deep and complex mathematics is involved in this process. In particular, the problem of deciding the best route to send the message led Bell Laboratories mathematician Narindra Karmarkar to invent a revolutionary new optimization algorithm a few years ago.

But sophisticated mathematics requires sophisticated engineering in order to make it work. In 1976, AT&T engineers pioneered a new system to help perform the complex switching task, known as "out-of-band signaling." This used a quite separate communications channel outside the normal "bandwidth" of phone calls in order to obtain and transfer the necessary information.

It was very effective. Information could be flashed to and fro from one side of the country to the other in seconds, allowing the mathematics to perform its task. The time lapse between the completion of dialing and the phone ringing at the receiving end was reduced from 15–20 seconds to an amazing four to six seconds.

But the more complex something is, the harder it is to change it. New developments intended to make things even better can introduce unforeseen difficulties. This is what happened on January 15, not for the first time and doubtless not for the last.

Major software systems are always in the process of being revised. The switching system used by AT&T has already undergone 13 upgrades, and the out-of-band signaling system was under regular development to keep abreast of changes in internationally agreed communications standards.

In order to satisfy the most recent standard, CCS-7 ("Common Channel Signaling" standard number 7), AT&T engineers had to modify their software in order to allow for, in particular, the transmission of the telephone number of the caller along with the call.

Normally, any difficulties encountered in such developments can be identified and corrected before the new version is put into operation. But this time things happened otherwise. When the new software was loaded on Sunday afternoon, a small error developed and propagated through the network and rapidly brought the system to its knees.

It was, of course, a "human error." And human errors are something mankind has always had to live with. Only the scale of the possible consequences has changed. The very complexity of present-day transport and communications networks makes them particularly vulnerable to tiny malfunctions leading to extensive disruption.

But perhaps this is not much more than a new twist to an old problem. Instead of writing this article in Louisville as I had planned, I am sitting in front of a computer screen at a friend's house in Palo Alto.

Having eventually managed to make all my confirming telephone calls, I arrived at the airport to find that I could not after all travel to Kentucky that day. All flights in and out of the connecting airport of Chicago were canceled. No engineering problems. No software crashes. No systems failures. Just that familiar old adversary. Fog.

135

The irony of information

March 1, 1990

In January, a small computer error at an exchange in New Jersey led to a major failure of the vast US-wide AT&T telephone network [Chapter 134]. For nine hours, over half the long-distance telephone calls attempted in the US failed to get through.

Though the fault occurred on a public holiday, when most businesses were closed, the failure nevertheless caused nation-wide chaos in airline bookings, hotel reservations, car rentals, and was a major obstacle to journalists, telephone shoppers, and a host of people who depend upon the telephone for their livelihood or leisure activities.

The next day it led to the publication of hitherto unknown access numbers which allow AT&T subscribers to route their long distance calls via one of the two much smaller competitors, MCI and Sprint, at no extra charge. This may result in a permanent loss of revenue for AT&T, since when someone receives the "all lines busy" message, they might not "try again later" as requested, but try another company instead.

AT&T's president described the breakdown as the most serious network problem in the company's history. A large portion of the daily life of a powerful and wealthy nation with more than 200 million people was disrupted by a small error in a computer program—proof indeed that ours is an Age of Information.

But what exactly is information? With so much of the stuff around, you might think the answer is obvious, but just what is it that we get from reference books and newspapers, TV screens, the radio and telephone network?

A few moments reflection should convince you that you don't have the foggiest idea. Nor does anyone else, including those Silicon Valley whiz-kids who have made fortunes out of the information-technology business.

True, we can pass on and obtain information in various ways: by voice, sound, pictures, words, touch, smell, and so on. But these are just the things that carry the information. The medium is not the message.

In fact, information is like the Cheshire cat's grin: it's what is left when you take away the electrons, the printed letters on the page, or the physical sound waves in the air. You cannot see the information, nor feel, touch, hear, or smell it. But it is there. You can acquire it and act upon it. You might even fight to protect it.

Why is it that we cannot answer such a basic question? Perhaps it is because there is not yet a science of information to explain the technology of information.

This situation is by no means new. If you lived in the Iron Age and asked an ironsmith "What is iron?", what kind of answer would you get? Very likely your craftsman would point to various artifacts he had made and tell you that each of those was iron. But this is not what you want. What you want to know is what makes iron the substance it is, and not some other substance that may or may not look quite like iron.

The problem is that your Iron Age man has no frame of reference with which to understand what you are asking. To give an answer that would satisfy you, he would need to know about the molecular structure of matter—for that surely is the only way to give a precise definition of iron. But he doesn't even conceive of the possibility of such a theory!

Similarly, the difficulty in trying to define information lies in the absence of an underlying theory upon which to base an acceptable definition. Like the Iron Age man and his stock in trade, Information Age man can recognize and manipulate information, but is unable to define precisely what it is that is being recognized and manipulated.

Given the dramatic importance of information in today's society, the absence of a science of information has not gone unnoticed. Around the world, research groups are working hard to try to come up with the right theory. The next *Micromaths* will examine one of these attempts [Chapter 136].

136

Information overload
March 15, 1990

What is information? This was the question raised here on March 1 [Chapter 135]. The problem is that our successful information technology is not yet supported by an information science. As I suggested last time, the situation is like traveling back in time to the Iron Age and asking an expert iron-craftsman what iron is. Though Iron Age man had a good technology of iron, it wasn't until this century that the atomic theory of matter enabled us to explain what distinguished iron from any other substance, by specifying its atomic structure.

At major research centers around the world, scientists are currently searching for the Information Age equivalent of the atomic theory of matter: a science of information. My own investigations took me to one such center, the Center for the Study of Language and Information (CSLI) at Stanford University in California, from 1987 to 1989, and then again for a brief visit last January.

My involvement began while I was at Lancaster University with my participation in a university/industry collaborative research program supported by the Alvey Directorate. The aim was to develop a sophisticated programming language to handle information.

But again and again I kept coming up against the same problem: once you started to look closely enough, there did not seem to be such a thing as "information." There were sound waves, words, pictures, messages, lists, charts, tables, books, electric signals, electromagnetic waves, magnetic tapes and discs and what-have-you, all of which could convey or store information. But these were the carriers. What the devil were they carrying?

Physicists and chemists had made progress by thinking of matter as made up of elementary particles. At first these were taken to be atoms. Then, when subsequent investigations indicated that this was not quite right, of still more basic particles, the protons, electrons, neutrons, and so on. Still later, this picture had to be changed again to allow for evermore fundamental particles.

It is an oft-debated point as to whether matter is really this way or whether we have simply found a way of thinking about things that works well for the time being but will one day have to be abandoned. Such questions only become relevant when things start to go wrong. While the "particle approach" works, we stick with it.

So, in the absence of any better ideas, it seemed to me to be worth trying "particles" of information. I came up with the name "infons." At least then there would be something on the table to theorize about.

But how should these "infons" look? They could not be thought of as physical particles buzzing about in the air. Information is quite different from physical matter, so its "particles" should be different from the physicist's particles. All I was borrowing was the idea of regarding things in terms of basic "items." The rest had to come from looking at information as it is.

At this stage, things came to an abrupt stop until, faced with the advice from my financially beleaguered vice-chancellor at Lancaster that I should move elsewhere, I decided to accept an offer of a visiting professorship at Stanford. Thus I became involved in the work being done at CSLI, where one of the lines being pursued by the 100 or so researchers was a new subject called Situation Theory, invented by Jon Barwise and John Perry, the founding directors of the Stanford institute.

And there, sitting almost unnoticed in their emerging theory, I saw my "infons." They had not been thought of as particles until then, and their role in the Barwise and Perry theory had been minimal. But they had most of the features I thought any sensible "particle of information" should have. And so infons were born.

Like the elementary particles of physics, infons come in two forms, positive and nega-tive. But there any similarity ends. Infons turn out to be much more like the various number systems of mathematics, in particular the real and imaginary numbers. That is: they are "mathematical particles."

The million dollar question is: does the introduction of infons lead to anything new? It is too early to say, but there is some encouraging evidence, and there is considerable work being done in the area.

During my January trip to CSLI, I took part in a lively debate led by the British artificial intelligence researcher, Pat Hayes, of the Xerox Palo Alto Research Center (PARC) on whether it really helped to think about information in terms of infons. (Hayes thinks it does not.) A week later I took part in a similar discussion at the Theories of Partial Information conference held at the University of Texas at Austin.

So even if the Barwise and Perry Situation Theory, and my suggestion of "particles" of information, fails, it might lead to something better. And that is all any scientist can ask for.

137

Out for the count

April 12, 1990

Prime numbers have turned out to have a whole host of properties that dazzle the intellect and baffle the imagination. A prime number is any whole number greater than 1 that is not evenly divisible by any number other than itself and 1. For instance, 2, 3, 5, 7, 11, 13, 17, 19 are the prime numbers up to 20. Though the proportion of whole numbers that are prime thins out as you proceed up through the numbers, Euclid proved in 350 B.C. that the primes are infinite in number.

Some of the most startling observations about prime numbers have resisted all attempts at a formal proof. A well-known example is the Goldbach Conjecture, first proposed in 1742; every even number greater than 2 is a sum of two primes. This is easy to check for the first few cases: $4 = 2 + 2$, $6 = 3 + 3$, $8 = 3 + 5$, and so on. And a fairly simple program will get your home micro to check the first thousand or so cases. You may even discover that the result is valid for all even numbers up to a billion. But for all that, no one has any idea whether or not the Goldbach Conjecture is really true or not.

Less well known is the Gilbreath Conjecture. This concerns the table shown. You start out by listing all the primes, as far as you want to go. Then on the line below you write down the differences between each successive pair of primes (subtract the smaller from the larger of each pair). Then on the third line you write down the differences between each successive pair of numbers on the second line. And so on.

Now for the curious part. No matter how far down you go, every single line starts with the number 1. You can check it by first trying a few values by hand, then teach your computer to look at a few hundred more.

But will you always get a sequence starting with a 1, no matter how far you carry the calculation? If so, there surely has to be some reason for that, and there ought to be a proper mathematical proof of the fact. But no one has been able to come up with an explanation.

Not all questions about prime numbers remain unanswered, however. Indeed, the February edition of *Mathematics Magazine* (published by the Mathematical Association of America) contains a new result about prime numbers discovered by Stanford mathematician and computer scientist Donald Knuth.

Knuth has found two 17-digit number, M and N, having no common factors, such that the following Fibonacci-like sequence contains *no* prime numbers: Start out by writing down M and N. Then write down their sum, $M + N$. Then continue the sequence by repeatedly writing down the sum of the final two numbers in the list to give the next number to add to the list.

Some years ago, the American mathematician Ronald Graham discovered two very large numbers, M and N, with this prime property (Graham's numbers had 33 and 34 digits respectively). But Knuth's new numbers are much smaller than this. The discovery of the new numbers involves a clever blend of modern computer power and some slick mathematical techniques for working with very large numbers—the essence of nineties' mathematics, and definitely a sign of things to come.

```
2  3  5  7 11 13 17 19 23 29 31 37 41 43 47 53 59 61 67 71
 1  2  2  4  2  4  2  4  6  2  6  4  2  4  6  6  2  6  4
  1  0  2  2  2  2  2  4  4  2  2  2  2  0  4  4  2
   1  2  0  0  0  0  2  0  2  0  0  0  2  4  0  2
    1  2  0  0  0  0  2  2  2  0  0  2  2  4  2
     1  2  0  0  0  2  0  0  0  2  0  2  0  2  2
      1  2  0  0  2  2  0  0  2  2  2  2  2  0
       1  2  0  2  0  2  0  2  0  0  0  0  2
        1  2  2  2  2  2  2  2  0  0  0  2
         1  0  0  0  0  0  0  2  0  0  2
          1  0  0  0  0  0  2  2  0  2
           1  0  0  0  0  2  0  2  2
            1  0  0  0  2  2  2  0
             1  0  0  2  0  0  2
              1  0  2  2  0  2
               1  2  0  2  2
                1  2  2  0
                 1  0  2
                  1  2
                   1
```

FIGURE 38. Prime differences. The Gilbreath Conjecture.

138

Odds on a perfectly odd number

April 26, 1990

Are there are odd perfect numbers? The question is easy to state, intriguing, and has a long history, but it remains unanswered. What we do now know, thanks to recent work by Richard Brent and Graeme Cohen, is that any odd perfect number would have to have at least 300 digits, which puts them firmly into the realm of computer mathematics.

The notion of a number being "perfect" goes back to Pythagoras in the 6th century B.C. He noticed that the number 6 has the curious property that it is equal to the sum of its own divisors (other than the number 6 itself): $6 = 1 + 2 + 3$. The next number after 6 with this property is 28. The divisors of 28 are 1, 2, 4, 7, 14, and $28 = 1+2+4+7+14$. Pythagoras called such numbers "perfect."

Since Pythagoras's time, other perfect numbers have been discovered. In the 1st century A.D., Nichomachus published the four known perfect numbers: 6, 28, 496, and 8,128. This list led to speculation that all perfect numbers are even, that they end alternately in 6 or 8, and that the nth perfect number has n digits.

The discovery of the next two perfect numbers put paid to all but the first of these conjectures: the fifth perfect number is the 8-digit 33,550,336, and the sixth is 8,589,869,056. But at once a fourth conjecture springs to mind: perfect numbers soon become very large. This is indeed the case, and the largest of the 30 perfect numbers now known has more than a hundred thousand digits.

The most elegant mathematical result concerning perfect numbers is undoubtedly Leonhard Euler's 18th century discovery that the even perfect numbers are of the form $2^{N-1}(2^N - 1)$, where N is such that the number $2^N - 1$ is prime. This has been the key to all the perfect numbers found since Euler's time. Now computers are used to find numbers N for which the number $2^N - 1$ is prime. The most recent discovery of this kind was made in 1986 using a Cray supercomputer [Chapter 63].

Other interesting facts about even perfect numbers include the following. First, all even perfect numbers do end in either 6 or 8, though not alternately. Second, every even perfect number is "triangular," which is to say it can be represented as the number of balls arranged to form an equilateral triangle. Third, every even perfect number beyond 6 is a sum of consecutive odd cubes. For example:

$$28 = 1^3 + 3^3$$

$$496 = 1^3 + 3^3 + 5^3 + 7^3.$$

But while progress was being made concerning the even perfect numbers, there remained a fundamental question about odd perfect numbers: are there any? All the available evidence suggests that there are no odd perfect numbers. But that evidence is pretty thin, amounting to little more than the use of a combination of mathematical and computational techniques to establish a minimum size that any odd perfect number would have to be.

In 1957, Kanold showed that an odd perfect number would have to have at least 20 digits. In 1973, Tuckerman pushed that limit to 36 digits, and in the same year Hagis capped that with 50 digits. Then in 1976, Buxton, Elmore and Stubblefield announced that they had shown there could be no odd per-

fect numbers with fewer than 200 digits, but the announcement was never followed up by a published proof, and some skepticism was expressed about its reliability.

Now, as a result of a 3,000-line, computer-generated proof, produced by means of a Pascal program running on a DEC VAX mini, Brent and Cohen have settled the matter definitively. There are no odd perfect numbers with fewer than 300 digits. But what of the possibility of there being odd perfect numbers with more than 300 digits? No one knows.

139

World's most wanted number

August 16, 1990

The announcement that a particularly notorious 155-digit number had been factored was yet another milestone in the march towards computational territory that only five years ago was thought to be a lifetime away.

The number is F_9, the ninth Fermat number. For some years it has been at the top of a widely circulated list of "most wanted" numbers that were known not to be prime but which had not been factored.

F_9 is not a prime number, so it must be a product of prime numbers. But apart from one 7-digit prime factor that had been discovered, nothing else was known about F_9's prime factorization.

The Fermat numbers are named after the 17th-century French mathematician Pierre de Fermat, who first investigated them. The Nth Fermat number, F_N, is obtained by first raising 2 to the power N, then raising 2 to that power, and then adding 1. Thus F_1 equals 5, F_2 equals 17, and F_3 equals 257. The next one is 65,537, so you can see that the Fermat numbers grow large pretty rapidly. Try working out the fifth one yourself.

Particular interest in Fermat numbers surrounds the question as to which are prime and which are not, following Fermat's original conjecture that they are all prime.

Although the four Fermat numbers small enough to handle using pencil and paper are indeed all prime, it has long been known that Fermat's conjecture was way off the mark. Indeed, computer techniques have failed to turn up any prime Fermat numbers apart from the first four.

Further interest in the primality of Fermat numbers stems from an intriguing connection, discovered by Gauss, between prime Fermat numbers and the classical Greek problem of ruler-and-compasses construction. But most present-day interest centers around the development of sophisticated algorithms to try to factorize Fermat numbers.

The discovery a few years ago of an ingenious cryptographic system (now widely used in the commercial world) that depended for its security on the difficulty of factoring large numbers—those with 100 digits or more—has led to the growth of a substantial computer factorization community. They spend their time, and millions of dollars-worth of computer time, searching for ever more efficient methods of factoring large numbers. Simply running through all the possible factors (trial and error) is impractical for all numbers having more than 20 digits. More sophisticated mathematical techniques have to be used. Because they are easy to describe, and yet are very large, numbers such as the Fermat numbers provide an excellent test-bed for any new factorization method.

One ingenious method, known as the number field sieve, was invented by the British mathematician John Pollard some years ago (and explained in *Micromaths* in October 1985 [Chapter 59]). This approach was used in factorizing F_9.

Credit for the actual computation goes to Mark Manasse of DEC and Arjen Lenstra, of Bellcore, the research lab run jointly by the American Bell operating companies, together with hundreds of other mathematicians spread around the globe, who kept in touch with each other using electronic mail.

Like the better known and closely related quadratic sieve, the number field sieve works by first breaking the task of factoring a single large number into a huge set of much smaller factoring problems. In the recent work, these were farmed out to a host of individuals around the world, who each worked on their own computers to produce the factors.

After two months, enough smaller factorizations had been completed to enable Manasse and Lenstra to try to piece these results together in the right way to obtain the master factorization.

This task was performed in a mammoth three-hour computation on a Connection Machine, a massively parallel supercomputer, at the Supercomputer-Computational Research Institute at Florida State University.

The result was a factorization of F_9 into a product of three prime numbers, the 7-digit prime already mentioned and two further primes, one having 49 digits, the other 99.

An intriguing aspect of the number field sieve method used in the latest success, is that the smaller factoring problems are not done using ordinary numbers, as is the case with the quadratic sieve, but with "algebraic integers"—roots of certain kinds of polynomials. This approach works particularly well for Fermat numbers, though not for other kinds of factorization.

Although the F_9 factorization sets a record it does not suggest an imminent breakdown of the data security system that depends on the difficulty of factoring. Indeed, the effort needed to perform the feat simply emphasizes just how difficult the factorization problem is.

140

A yen for teamwork

December 12, 1991

Japan is a world leader in hardware manufacturing, but its position in software is far less certain. The gap can be understood in terms of the Japanese emphasis on conformity and teamwork as opposed to the individualistic approach of Americans, who lead the world in software development.

But what happens when software development starts to move towards the production of tools designed for teamwork? Will the Japanese approach start to pay off at the computer keyboard? Well, that was the suggestion put forward by one of the Japanese hosts at IMSA '91, the International Symposium on New Models for Software Architecture, at Waseda University, Tokyo, last month.

The symposium marked the start of an eight-year project, established in 1990 by Japan's Ministry of International Trade and Industry (MITI). This is aimed at developing new models for software to support cooperative work—work carried out by large research and development teams—as opposed to the single-user model on which today's PCs and work-stations are based.

The Japanese learned from their Fifth Generation Project a decade ago that the scale of the problems involved in developing new software needs an international approach. Accordingly, the New Models for Software Architecture project (NMSA) is intended to be an international collaboration.

The development of computer tools to support teamwork is not new. Probably the first purpose built, distributed, office support system was developed by workers at Xerox's famous Palo Alto Research Center, California, in the late 1970s. Their networked system of Alto (later Star) computers, pioneered the windows-icons-mouse-pointer (WIMP) interface—later made popular by the Apple Macintosh—together with shared printers and information storage devices.

"Personal distributed computing" was the name Xerox attached to this kind of integrated system. Following a conference held at the University of Texas at Austin in 1986, the concept originated at PARC developed into the field now known as Computer Supported Co-operative Work (CSCW). It's a concept that has had a long wait for acceptance.

For its first 30 years, computing was driven by technological developments. Computer science was seen first as a branch of mathematics, and later as an engineering discipline. The results were systems that showed technical brilliance, but their use was limited to that small number of people with the desire, the aptitude, and the training to use them.

Meanwhile, PARC was busy recruiting not mathematicians and computer scientists but linguists, philosophers, cognitive psychologists and social scientists. They looked at how the computer could be designed to suit the user, not the other way around. That is why PARC's ideas, though not Xerox's products, are starting to dominate the computer world. Obviously, if you want to design a computer system to support teamwork, you must start by looking at the way teams already work. This is the CSCW approach to software development, and it results in a drastic change in the computer science hierarchy. Mathematicians and engineers are having to share their preeminence with all those "humanities and social science types."

However, it was precisely these disciplines that were savaged in the restructuring of the UK university system, which was partly aimed at promoting the development of information technology. The results were predictable. As one of the British participants in the IMSA '91 conference remarked, the UK's Alvey Programme led to the production of some technically sophisticated systems which were of limited use in the field.

As the Japanese now acknowledge, truly useful developments in information support systems will depend on work that is not only international but also interdisciplinary. The work will involve not just computer science and software engineering but all of the disciplines mentioned above. The successful computer scientist of tomorrow will be both mathematician and sociologist, engineer and philosopher, linguist and psychologist.

141

Math gang makes Fermat prime suspect
May 22, 1992

The discovery of ever larger prime numbers gets the headlines these days, but for the professional mathematician such announcements evoke barely a yawn. After all, back in 350 B.C. the great Greek mathematician Euclid proved there are infinitely many prime numbers, so it comes as no surprise when a new, larger one is found.

(A prime number is one that can only be evenly divided by itself, and 1. Thus, 2, 3, 5, 7, 11, 13, 17, and 19 are the prime numbers less than 20. Prime numbers become rare the higher you go, though by Euclid's theorem, they never peter out.)

Far more significant was the discovery earlier this year that there are infinitely many Carmichael numbers. Credit for this discovery goes to three mathematicians at the University of Georgia: Englishman Andrew Granville, formerly of Cambridge University, and Americans Carl Pomerance and Red Alford. Their discovery represents the final nail in the coffin for a putative test for prime numbers that goes back to the great French amateur mathematician Pierre de Fermat.

On October 18, 1640, in a letter to his colleague Frenicle, Fermat noted that if P is any prime number, then for any number N that you choose, P exactly divides $N^P - N$. This was a truly remarkable observation of the "think of a number variety." For example, take the prime number 3. Then, pick any num-

ber N you like; raise N to the power 3; then subtract N. The answer will always be divisible by 3. Suppose you pick $N = 8$. Raising 8 to the power 3, you get 512. Subtract 8 from 512 to give 504. This number 504 is now divisible by 3.

Or take $N = 35$. Then $35^3 = 42875$. Subtract 35 and you get 42840. Again, this number can be evenly divided by 3. Known as Fermat's Little Theorem (to distinguish it from the famous Fermat's Last Theorem, which to this day remains unproved), this fact lies behind many common card tricks where a conjuror will "predict" a chosen card. To mathematicians, however the main interest lies in its ability to test for primality. Or to be more precise, to test for non-primality.

Primality testing became big business about a decade ago when a highly secure form of data encryption was developed that makes use of large prime numbers, prime numbers having a hundred or so digits. For a number P with a hundred digits, testing to see if P is prime by looking for numbers that divide into P could take billions of centuries, even using the world's fastest computers. So mathematicians use other methods.

A particularly simple method uses the Fermat Little Theorem. To test if a number P is prime, see if P divides into $2^P - 2$. If it does not, you know P cannot be prime. But what happens if P does divide $2^P - 2$? Well, the chances are that P is prime, but you cannot be sure. The problem is, though P always divides $2^P - 2$ if P is prime, there are also some non-prime values for P that have this property; for example, 341 does, and yet $341 = 11 \times 31$, so this number is not prime.

How about trying a different number, say 3, working out $3^P - 3$? Or 4, or 5, or 6, or 7, and so on; maybe one of these will work? Unfortunately, there are some non-prime numbers P for which, whatever number N you take, $N^P - N$ is divisible by P. The smallest such is 561 ($= 3 \times 11 \times 17$). These numbers P are called *Carmichael numbers* after the man who, in 1910, found that 561 had this property.

From Carmichael's time, the question was, just how many such numbers are there? There are none other than 561 below 1,000, and only six more below 10,000. There are just 43 less than a million.

Using high powered computers, a number of mathematicians pushed the search further and further, with Pinch at Cambridge finding 105,212 Carmichael numbers less than 10^{15} earlier this year. With the discovery of so many numbers of this kind, it soon be-

came clear that the Fermat Little Theorem was clearly not suitable to use as a genuine and reliable test for primality, and in fact the smart money was on there being infinitely many such numbers, just as there are infinitely many prime numbers.

Well, the smart money turned out to be very smart. Using some highly sophisticated mathematical techniques, what the University of Georgia team recently discovered is the following strange-seeming fact. There is some number K such that for any number X bigger than K, there are more than X raised to the power of 2/7 Carmichael numbers less than X. Though the Georgia three have no idea how big this number K is, their mathematics guarantees such a number exists. Then, since you can make the number X raised to the power 2/7 as big as you please by choosing X big enough, it follows that the number of Carmichael numbers must be infinite.

Which might be the end of the story. But remember, Euclid's 2,000-year-old proof that there are infinitely many primes turned out to be just the start of a long and of late expensive computational odyssey.

142

Record primes

Added July 1993

In March 1992, it was announced that yet another Mersenne prime had been discovered, the 32nd. A Cray-2 computer in England proved that the 227,932-digit Mersenne number $2^{756839} - 1$ is prime. Commenting on the discovery in the Usenet, the electronic bulletin board system used by scientists to communicate with each other in a public forum, Robert Silverman of the US-based Mitre Corporation added that it was clear that the new Mersenne prime was the next one in magnitude. Indeed, Silverman observed, there remain huge stretches of Mersenne exponents that have never been checked. The range of exponents from 170,000 to 216,091 has not been completely checked, and the range from 360,000 to 430,000 has been only partially examined. The range from 430,000 to 520,000 has been completely checked, but the range from 520,000 up to 750,000 has been examined only sparsely.

143

Fermat's Last Theorem, a theorem at last?

Added April 1994

On June 23, during the course of a series of three lectures at a conference held at the Newton Institute in Cambridge, British mathematician Dr. Andrew Wiles, of Princeton University, presented a proof of a deep new result that might well imply Fermat's Last Theorem. More precisely, Wiles sketched a proof of the Shimura–Taniyama–Weil conjecture for a certain class of semi-stable elliptic curves. As Kenneth Ribet, of the University of California at Berkeley, showed some years ago, Fermat's Last Theorem is a corollary of this conjecture when it is applied to *all* semi-stable curves. The only remaining question is, do the curves for which Wiles' argument work include those required to prove Fermat's Last Theorem? The answer may be yes, but so far, no one knows for certain. [See Chapter 91.]

Wiles' approach makes use of geometric structure, and depends, in particular, on an amazing connection, established during the last decade, between the Last Theorem and the theory of elliptic curves. Elliptic curves are determined by equations of the form

$$y^2 = x^3 + ax + b$$

where a and b are integers. (Note that an ellipse is not an elliptic curve. It is the problem of finding the length of part of an ellipse that gives rise to an "elliptic curve," and hence the name.)

The study of elliptic curves has proved to be a rich, and extremely fruitful, area of research in number theory, and, indeed, one of the most powerful known methods to factor large integers depends on this theory, giving the theory strategic significance in the murky, real-world activity of cryptography.

In 1955, the Japanese mathematician Yukata Taniyama proposed that there should be a connection between elliptic curves and another well-understood class of curves, known as modular curves. According to Taniyama, there should be a connection between any given elliptic curve and a modular curve, and this connection should 'control' many of the properties of the initial curve.

Taniyama's conjecture was made more precise in 1968 by André Weil, who showed how to determine the exact modular curve that should be connected to a given elliptic curve, and, in 1971, Goro Shimura demonstrated that this works for a very special class of equations. Taniyama's proposal became known as the Shimura–Taniyama–Weil Conjecture.

But so far, there was no known connection between this very abstract conjecture and Fermat's Last Theorem. Things changed dramatically in 1986, when Gerhard Frey, from Saarbrucken, discovered a most surprising and innovative link between the two.

What he realized was that if there are numbers a, b, c, n such that

$$c^n = a^n + b^n,$$

then it seemed unlikely that one could understand the elliptic curve given by the equation

$$y^2 = x(x - a^n)(x + b^n)$$

in the way proposed by Taniyama. [See Chapter 91.]

Following an appropriate re-formulation by Jean-Pierre Serre, Kenneth Ribet proved conclusively that the existence of a counterexample to the Last Theorem would in fact lead to the existence of a elliptic curve which could not be modular, and hence would contradict the Shimura–Taniyama–Weil Conjecture.

This gave mathematicians a definite structure to work with. Whereas there is no reason to believe the statement of the Last Theorem, the Shimura–Taniyama–Weil conjecture concerned geometric objects about which a great deal was known, sufficient structure being understood to give good reason to believe the result, and indeed to suggest a way of setting about finding a proof. At least, Andrew Wiles saw a way to proceed.

For the next seven years, Wiles concentrated all his efforts on trying to find a way to make his idea work. Using and developing powerful new methods of Barry Mazur, Matthias Flach, Victor Kolyvagin, and others, he eventually succeeded in establishing the Shimura–Taniyama–Weil conjecture for an important class of elliptic curves (those with squarefree 'conductor'), which may include those relevant to proving Fermat's Last Theorem.

In fact, at first Wiles was sure that his proof did imply the Last Theorem, and as a result his June announcement was of a proof of Fermat's longstanding teaser. But a few months later, he realized that there was a "gap" in his proof. That gap remains to this day, but it seems that we are one step further towards the final resolution of the Last Theorem. As always, mathematics moves forward. And, as always, there is still another step to be made.

Topic Index

(Chapter numbers are in boldface, page numbers in lightface.)